Decomposition of Multivariate Probabilities

This is Volume 29 in
PROBABILITY AND MATHEMATICAL STATISTICS
A Series of Monographs and Textbooks

Editors: Z. W. Birnbaum and E. Lukacs

A complete list of titles in this series appears at the end of this volume.

Decomposition of Multivariate Probabilities

ROGER CUPPENS

U.E.R. de Mathématique
Université Paul Sabatier
Toulouse, France

ACADEMIC PRESS New York San Francisco London 1975
A Subsidiary of Harcourt Brace Jovanovich, Publishers

ACADEMIC PRESS, INC.
111 Fifth Avenue, New York, New York 10003

United Kingdom Edition published by
ACADEMIC PRESS, INC. (LONDON) LTD.
24/28 Oval Road, London NW1

Library of Congress Cataloging in Publication Data

Cuppens, Roger.
Decomposition of multivariate probabilities.

(Probability and mathematical statistics series;)
Includes bibliographical references and indexes.
1. Probabilities. 2. Decomposition (Mathematics)
3. Multivariate analysis. I. Title.
QA273.C86 519.5′3 74-10212
ISBN 0-12-199450-3

PRINTED IN THE UNITED STATES OF AMERICA

Contents

Chapter 8 **Infinitely Divisible Probabilities with Countable Poisson Spectrum**

Chapter 9 **α-Decomposition**

Appendix A **Some Results of Function Theory**

Appendix B **Exponentials of Polynomials and Functions**

Preface

Since the appearance in 1960 of the book "Characteristic Functions" by Eugene Lukacs, it has become well-known that characteristic functions of probability laws, which were originally created as a tool for limit theorems in probability theory and which have been applied successfully to many problems in probability theory and mathematical statistics, have an intrinsic mathematical interest. Many great advances have been made in this area since that date.

But the book by Lukacs and other books on this subject are concerned with characteristic functions of probabilities defined on the real line, and the important problem of multivariate characteristic functions has not been studied. The present book is devoted to this problem.

The results on multivariate characteristic functions are historically of two types: the "classical" results (which appeared essentially before 1940) and the recent ones. Most of the recent work has been published in Russian or French journals. Since these are not easily accessible, we think that a complete report will be beneficial.

After a first chapter devoted to some useful results on measures and integrals, we give the classical theory in Chapter 2. We have enlarged the problem and studied the Fourier–Stieltjes transforms of signed measures. This very useful study is no more difficult than the classical one.

In Chapter 3 we study the notion of analytic characteristic functions. For multivariate characteristic functions, this study is quite recent. At the end of this chapter, we again step beyond the limits of characteristic functions by giving the multivariate extension of the well-known Paley–Wiener theorem on functions that are entire, of exponential type and square-integrable.

In Chapter 4 we begin with decomposition problems. The theory of infinitely divisible probabilities and the classical results of Hinčin are given. Again, we have tried to enlarge the problem and give an approach slightly different from the classical one. The decompositions of analytic characteristic functions are studied in Chapter 5.

The next three chapters are devoted to the important problem of the description of the class I_0^n of n-variate probabilities without indecomposable factors. We give the present status of this problem which is not yet completely solved. Some of the results given here are new.

Finally, in Chapter 9, we study the problem of α-decompositions of multivariate characteristic functions, while in an appendix we give some nonclassical function theory results which are useful for the study of multivariate characteristic functions. In another appendix we state the results due to Lévy on the expansion in power series of exponentials of polynomials.

Following the ideas of Lukacs, we have not studied the possible applications of characteristic functions. In particular, the problem of stability of decompositions and the applications to limit theorems or characterization problems are outside the scope of this book.

We have not given proofs of the results on univariate characteristic functions when these can be found in Lukacs's book. All the other proofs are complete. In the bibliographical notes we have tried to sketch the development of the theory, and the bibliography has been determined accordingly; a complete bibliography would contain many hundreds of items, but we have listed only the most significant works, either historically or mathematically, concerning the mathematical theory of characteristic functions.

I should like to express my thanks to Professor E. Lukacs for including this work in the series edited by him and Professor Birnbaum, and to Mr. and Mrs. B. Rousseau and to Mrs. N. Macle for their help during the preparation of the manuscript.

Notation

If A and B are two subsets of R^n, then $A \cup B$ and $A \cap B$ denote, respectively, the union and the intersection of A and B, and A^c denotes the complementary set of A and

$$A \backslash B = A \cap B^c = \{x \in R^n : x \in A, x \notin B\}.$$

If A is a subset of R^n, we denote by $\mathring{A}$, $\bar{A}$, and ∂A, respectively, the interior, the closure, and the boundary of A.

If A is a subset of R^n and $j \in R$, we denote by jA the set

$$jA = \begin{cases} \{x \in R^n : j^{-1}x \in A\} & \text{if} \quad j \neq 0, \\ \{0\} & \text{if} \quad j = 0. \end{cases}$$

In particular, $-A$ is the symmetric set of A with respect to the origin. If A and B are two subsets of R^n, $A + B$ is the vectorial sum of these two sets

$$A + B = \{x \in R^n \mid \exists (y, z) \in A \times B : x = y + z\}.$$

Instead of $A + (-B)$, $A + \{m\}$, and $A + \{-m\}$, we write $A - B$, $A + m$, and $A - m$, respectively. If E_j is a subspace of R^n $(j = 1, \ldots, m)$

$$A = E_1 \oplus \cdots \oplus E_m$$

means that

$$A = E_1 + \cdots + E_m$$

and any $x \in A$ admits a unique representation

$$x = \sum_{j=1}^{m} x_j$$

where $x_j \in E_j$ $(j = 1, \ldots, m)$.

If $j \in N$, we define inductively $(j)A$ by

$$(0)A = \{0\}, \qquad (1)A = A,$$
$$(j)A = (j-1)A + A, \qquad \text{if } j > 1,$$

and $(N)A$ by

$$(N)A = \bigcup_{j=0}^{\infty} (j)A.$$

$(N)A$ is the set of all the linear combinations of elements of A with nonnegative integer coefficients. More generally, we denote by $(Z)A$ (resp. $(Q^+)A$, $(Q)A$) the set of all the linear combinations of elements of A with integer (resp. nonnegative rational, rational) coefficients.

We denote by e_j the element of R^n (or C^n)

$$e_j = (0, \ldots, 0, 1, 0, \ldots, 0)$$

where the jth coordinate is 1 and the others are 0. $\{e_1, \ldots, e_n\}$ is the canonical basis of R^n (or C^n).

If $z = (z_1, \ldots, z_n) \in C^n$, we define $\operatorname{Re} z$, $\operatorname{Im} z$, $|z|$ by

$$\operatorname{Re} z = (\operatorname{Re} z_1, \ldots, \operatorname{Re} z_n),$$
$$\operatorname{Im} z = (\operatorname{Im} z_1, \ldots, \operatorname{Im} z_n),$$
$$|z| = (|z_1|, \ldots, |z_n|).$$

If A and B are two subsets of R^n, then

$$A + iB = \{z \in C^n : \operatorname{Re} z \in A, \operatorname{Im} z \in B\}.$$

If $x = (x_1, \ldots, x_n) \in C^n$ and $y = (y_1, \ldots, y_n) \in C^n$, then (x, y) means the scalar product of x and y

$$(x, y) = \sum_{j=1}^{n} x_j \overline{y_j}$$

and $\|x\|$ means the norm of x

$$\|x\| = \sqrt{(x, x)}$$

(it must be noted that $|x| \in R^n$ while $\|x\| \in R$). If $(x, y) = 0$, x and y are orthogonal and we denote this by $x \perp y$. If E is a subspace of C^n, then

$$E^\perp = \{x \in C^n : x \perp y \qquad \text{for any } y \in E\}$$

is the orthogonal subspace of E. In particular, if E is the one-dimensional subspace generated by $\theta \neq 0$

$$E = \{x \in C^n : x = \lambda\theta, \lambda \in C\},$$

then we write $\theta^{\perp}$ instead of $E^{\perp}$ (all the preceding notations are also valid for R^n).

If $x = (x_1, \ldots, x_n) \in R^n$ and $y = (y_1, \ldots, y_n) \in R^n$, then we write $x \leqq y$ (resp. $x < y$) if $x_j \leqq y_j$ (resp. $x_j < y_j$) ($j = 1, \ldots, n$).

If $k = (k_1, \ldots, k_n) \in N^n$, then

$$k! = k_1! \cdots k_n!$$

and if $x = (x_1, \ldots, x_n) \in R^n$, $\alpha = (\alpha_1, \ldots, \alpha_n) \in R^n$, then

$$x^{\alpha} = x_1^{\alpha_1} \cdots x_n^{\alpha_n}$$

if the right-hand side of this equality is defined.

If f is a function defined on R^n with complex values and if $\alpha = (\alpha_1, \ldots, \alpha_n) \in N^n$, $m \in R^n$, we define $D^{\alpha} f(m)$ by

$$D^{\alpha} f(m) = \frac{\partial^{\alpha_1 + \cdots + \alpha_n} f}{\partial x_1^{\alpha_1} \cdots \partial x_n^{\alpha_n}}(m)$$

if this derivative exists.

Let E be some subspace of R^n. If $x \in R^n$ and if f is a function defined on R^n, we denote by $\mathrm{pr}_E\, x$ the projection of x on E and by $\mathrm{pr}_E\, f$ the function defined by

$$\mathrm{pr}_E\, f(x) = f(\mathrm{pr}_E\, x)$$

for any $x \in R^n$. ($\mathrm{pr}_E\, f$ is the composition of pr_E and f.) If E is the one-dimensional subspace generated by $\theta \neq 0$, then we write pr_{θ} instead of pr_E.

List of Symbols†

† Numbers refer to the pages where each symbol has been initially defined.

Chapter 1

Measures and Integrals

In this chapter, we recall some definitions and results on measures and Lebesgue integrals which will be useful. We do not attempt to be complete; in particular, the results are not given in their full generality, but in the form used in the following chapters. For terminology, we follow Halmos (1950).

1.1 MEASURES

We define as μ an *n-variate measure* if μ is a function defined on the class $\mathscr{B}_n$ of all the Borel sets of R^n with values in $[0, +\infty]$ and satisfying

$$\mu(\varnothing) = 0$$

and

$$\mu\left(\bigcup_{j=1}^{\infty} B_j\right) = \sum_{j=1}^{\infty} \mu(B_j)$$

for any sequence $\{B_j\}$ of disjoint Borel sets ($B_j \cap B_k = \varnothing$ for $j \neq k$). We define p as an *n-variate probability* if p is an n-variate measure satisfying

$$p(R^n) = 1.$$

We denote by $\mathscr{P}_n$ the set of all the n-variate probabilities.

Example 1 The measure μ defined by

$$\mu(B) = 0$$

for any $B \in \mathscr{B}_n$ is the measure 0.

Example 2 For any $\alpha \in R^n$, we can define a probability δ_α by

$$\delta_\alpha(B) = \begin{cases} 0 & \text{if } \alpha \notin B, \\ 1 & \text{if } \alpha \in B. \end{cases}$$

This probability is the degenerate probability (Dirac measure) at the point α.

Since a measure is a function with values in $[0, +\infty]$, we define the sum $\mu = \mu_1 + \mu_2$ of two measures μ_1 and μ_2 as

$$\mu(B) = \mu_1(B) + \mu_2(B)$$

and the product $\nu = \alpha\mu$ of a measure μ by a nonnegative constant α as

$$\nu(B) = \alpha\mu(B)$$

for any $B \in \mathscr{B}_n$.

An n-variate measure μ is finite if

$$\mu(R^n) = \|\mu\| < +\infty.$$

Then

$$0 \le \mu(B) \le \|\mu\|$$

for any $B \in \mathscr{B}_n$. If μ is a finite measure, $\mu \ne 0$, there exists a unique probability p such that

$$\mu = \|\mu\| \cdot p.$$

An n-variate measure μ is σ-finite if there exists a sequence $\{B_j\}$ of Borel sets such that

$$R^n = \bigcup_{j=1}^{\infty} B_j,$$

$$\mu(B_j) < +\infty \qquad (j = 1, 2, \ldots).$$

Any finite measure is σ-finite.

An important theorem is the following one.

Theorem 1.1.1 If $\mathscr{R} \subset \mathscr{B}_n$ is a ring such that the σ-ring generated by $\mathscr{R}$ is $\mathscr{B}_n$ and if μ is a σ-finite measure on $\mathscr{R}$, there exists a unique n-variate measure $\bar{\mu}$ (the extension of μ) such that

$$\mu(B) = \bar{\mu}(B)$$

for any $B \in \mathscr{R}$, and $\bar{\mu}$ is σ-finite.

Proof This is a direct consequence of the extension theorem [see Halmos (1950, p. 54)].

If μ is a finite measure and if we define F_μ by

$$F_\mu(x) = \mu(\{y : y < x\}), \tag{1.1.1}$$

F_μ is a function defined on R^n with nonnegative real values and having the following properties:

(a) F_μ is nondecreasing [$F_\mu(x_1) \leq F_\mu(x_2)$ if $x_1 \leq x_2$];
(b) $\lim_{x_j \to -\infty} F_\mu(x_1, \ldots, x_n) = 0$ $(j = 1, \ldots, n)$;
(c) $\lim_{x_1 \to +\infty, \ldots, x_n \to +\infty} F_\mu(x_1, \ldots, x_n) = \|\mu\|$;
(d) $F_\mu(x) = \lim_{\xi \uparrow x} F_\mu(\xi)$.†

A function defined on R^n with nonnegative values and having properties (a)–(d) is called an *n-variate distribution function.* [If μ is an n-variate finite measure, the function F_μ defined by (1.1.1) is the distribution function of μ.]

Conversely, since the class of all the finite unions and intersections of sets of the kind $\{y \in R^n : y < x\}$ is a ring, if F is an n-variate distribution function, Theorem 1.1.1 implies the existence of a unique n-variate finite measure having F for distribution function.

1.2 INTEGRALS‡

If $B \in \mathscr{B}_n$, we denote by i_B the indicator of the set B

$$i_B(x) = \begin{cases} 0 & \text{if } x \notin B, \\ 1 & \text{if } x \in B, \end{cases}$$

and call *simple* a function f defined by

$$f = \sum_{j=1}^{m} \alpha_j i_{B_j}$$

where $B_j \in \mathscr{B}_n$, $\bigcup_{j=1}^m B_j = R^n$, $\alpha_j \in R$ $(j = 1, \ldots, m)$. If $\mu(B_j) < +\infty$ for $\alpha_j \neq 0$ $(j = 1, \ldots, m)$, then f is integrable and its integral is defined as

$$\int f(x)\mu(dx) = \sum_{\alpha_j \neq 0} \alpha_j \mu(B_j).$$

A sequence $\{f_j\}$ of integrable simple functions is *mean fundamental* if

$$\lim_{j\to\infty} \lim_{k\to\infty} \int |f_j(x) - f_k(x)| \mu(dx) = 0.$$

† $\xi = (\xi_1, \ldots, \xi_n) \uparrow x = (x_1, \ldots, x_n)$ means that $\xi_j \to x_j$, $\xi_j \leq x_j$ for $j = 1, \ldots, n$.
‡ For all the details of this section, see Halmos (1950, Chapter 5).

If f_j $(j = 1, 2, \ldots)$ and f are measurable functions defined on R^n with real values, the sequence $\{f_j\}$ converges in measure to f if

$$\lim_{j\to\infty} \mu(\{x : |f_j(x) - f(x)| > \varepsilon\}) = 0$$

for any $\varepsilon > 0$.

Then we say that a measurable function f is *integrable* if there exists a mean fundamental sequence $\{f_j\}$ of integrable simple functions which converges in measure to f. We define the integral $\int f(x)\mu(dx)$ of f by

$$\int f(x)\mu(dx) = \lim_{j\to\infty} \int f_j(x)\mu(dx).$$

(It can be shown that this limit exists and does not depend on the mean fundamental sequence $\{f_j\}$ converging to f.)

Now if f is a function defined on R^n with complex values, then $f = g + ih$ where g and h are functions defined on R^n with real values and we say that f is *integrable* if g and h are both integrable and we define the integral of f by

$$\int f(x)\mu(dx) = \int g(x)\mu(dx) + i \int h(x)\mu(dx).$$

If $B \in \mathscr{B}_n$, we define the integral of f on B with respect to the measure μ as

$$\int_B f(x)\mu(dx) = \int f(x)i_B(x)\mu(dx).$$

The main properties of this integral are the following.

(a) The integral is linear with respect to the function

$$\int (\alpha_1 f_1 + \alpha_2 f_2)(x)\mu(dx) = \alpha_1 \int f_1(x)\mu(dx) + \alpha_2 \int f_2(x)\mu(dx)$$

for any measurable functions f_1 and f_2 defined on R^n with complex values and any complex constants α_1 and α_2.

(b) The integral is linear with respect to the measure

$$\int f(x)(\alpha_1 \mu_1 + \alpha_2 \mu_2)(dx) = \alpha_1 \int f(x)\mu_1(dx) + \alpha_2 \int f(x)\mu_2(dx)$$

for any measures μ_1 and μ_2 and any nonnegative constants α_1 and α_2. (These two relations mean that if two of the integrals exist, then the third also exists and the three integrals satisfy the relation.)

(c) The integral is positive:

$$\int f(x)\mu(dx) \geq 0$$

for any nonnegative function f.

(d) Since the integral is linear and positive, we have the two inequalities

$$\left|\int f(x)\mu(dx)\right| \le \int |f(x)|\,\mu(dx),$$

$$\left|\int f(x)g(x)\mu(dx)\right|^2 \le \int |f(x)|^2\mu(dx)\cdot\int |g(x)|^2\mu(dx).$$

This last inequality is known as *Schwarz's inequality*.

Finally, we state the fundamental

Theorem 1.2.1 (*Lebesgue dominated convergence theorem*) If $\{f_j\}$ is a sequence of measurable functions which converges to f almost everywhere† and if there exists an integrable function g such that

$$|f_j(x)| \le g(x)$$

almost everywhere ($j = 1, 2, \ldots$), then f_j and f are integrable and

$$\lim_{j\to\infty}\int f_j(x)\mu(dx) = \int f(x)\mu(dx).$$

Proof See Halmos (1950, p. 110).

1.3 PRODUCT MEASURES

If μ_j is an n_j-variate σ-finite measure ($j = 1, 2$), from Theorem 1.1.1 we deduce the existence of a unique n-variate measure μ ($n = n_1 + n_2$) such that

$$\mu(A_1 \times A_2) = \mu_1(A_1)\cdot\mu_2(A_2)$$

for any $A_j \in \mathscr{B}_{n_j}$ ($j = 1, 2$). Also μ is σ-finite. We write then

$$\mu = \mu_1 \times \mu_2$$

and call this measure the *product* of μ_1 and μ_2.

For example, there exists a unique n-variate measure λ satisfying

$$\lambda(\{x \in R^n : a \le x \le b\}) = \prod_{j=1}^{n}(b_j - a_j)$$

for any $a = (a_1, \ldots, a_n) \in R^n$, $b = (b_1, \ldots, b_n) \in R^n$, $a \le b$. This measure is the n-variate Lebesgue measure.

† We recall that a property holds everywhere if the set A where the property does not hold is a null set [$\mu(A) = 0$].

A product measure is more precisely determined by the following result [see Halmos (1950, p. 144)].

Theorem 1.3.1 For any $B \in \mathscr{B}_n$

$$\int (\mu_1 \times \mu_2)(B) = \int \mu_2(B_x)\mu_1(dx) = \int \mu_1(B^y)\mu_2(dy)$$

where

$$B_x = \{y : (x, y) \in B\}, \qquad B^y = \{x : (x, y) \in B\}.$$

The fundamental result for product measures is

Theorem 1.3.2 (*Fubini's theorem*) Let f be a function defined on R^n with complex values. If f is integrable with respect to $\mu = \mu_1 \times \mu_2$, then

$$\int f(z)\mu(dz) = \int \left[\int f(x, y)\mu_2(dy)\right]\mu_1(dx)$$

$$= \int \left[\int f(x, y)\mu_1(dx)\right]\mu_2(dy),$$

this relation meaning that the integrals within brackets exist almost everywhere and are integrable.

Proof See Halmos (1950, p. 148).

1.4 SIGNED MEASURES

If μ_1 and μ_2 are n-variate measures one of which at least is finite, then the function $\mu = \mu_1 - \mu_2$ defined by

$$\mu(B) = \mu_1(B) - \mu_2(B)$$

for any $B \in \mathscr{B}_n$ is an n-variate signed measure. If μ_1 and μ_2 are both finite (resp. σ-finite), then the signed measure $\mu = \mu_1 - \mu_2$ is finite (resp. σ-finite).

If μ_1 and μ_2 are n-variate finite signed measures and α_1 and α_2 are real constants, we can define a finite signed measure $\mu = \alpha_1 \mu_1 + \alpha_2 \mu_2$ by

$$\mu(B) = \alpha_1 \mu_1(B) + \alpha_2 \mu_2(B)$$

for any $B \in \mathscr{B}_n$ and, with this definition, the set $\mathscr{M}_n$ of all the n-variate finite signed measures is a real linear space and is generated by $\mathscr{P}_n$.

If $A \in \mathscr{B}_n$ and $\mu \in \mathscr{M}_n$, μ is concentrated on A if

$$\mu(B \cap A) = \mu(B) \tag{1.4.1}$$

for any $B \in \mathscr{B}_n$. If we denote by $\mathscr{M}(A)$ the set of all the n-variate finite signed measures concentrated on A, then $\mathscr{M}(A)$ is a linear subspace of $\mathscr{M}_n$. Since (1.4.1) is equivalent to

$$\mu(B \cap A^c) = 0$$

for any $B \in \mathscr{B}_n$, it follows easily that if μ is concentrated on A_j $(j = 1, 2, \ldots)$, then μ is concentrated on $\bigcap_{j=1}^{\infty} A_j$. If $\mu \in \mathscr{M}_n$, we can prove the existence of a smallest closed set $S(\mu)$ on which μ is concentrated and of a smallest convex closed set $C(\mu)$ on which μ is concentrated. $S(\mu)$ is the support of μ and $C(\mu)$ is the convex support of μ. From a well-known result [see Halmos (1950, p. 66)], it follows that if μ is concentrated on A, there exists some $A_1 \in F_\sigma^n$ and some $A_2 \in G_\delta^n$ such that $A_1 \subset A \subset A_2$ and μ is concentrated on A_j $(j = 1, 2)$.† Without any loss of generality, we can therefore suppose that μ is concentrated on a set $A \in F_\sigma^n$.

Theorem 1.4.1 (*Hahn–Jordan decomposition theorem*) If $\mu \in \mathscr{M}_n$, there exist disjoint sets A^+ and A^- and finite measures μ^+ and μ^- such that μ^+ is concentrated on A^+, μ^- is concentrated on A^-, and $\mu = \mu^+ - \mu^-$. The measures μ^+ and μ^- are uniquely determined by μ.

Proof See Halmos (1950, pp. 121–123).

If we say that $\mu \le \nu$ if $\mu(B) \le \nu(B)$ for any $B \in \mathscr{B}_n$, the measures μ^+ and μ^- of the Hahn–Jordan decomposition theorem are minimal in the following sense.

Theorem 1.4.2 If μ_1 and μ_2 are finite measures satisfying $\mu = \mu_1 - \mu_2$, then

$$\mu^+ \le \mu_1, \qquad \mu^- \le \mu_2$$

where μ^+ and μ^- are the measures defined in the preceding theorem.

If μ^+ and μ^- are the measures defined in Theorem 1.4.1, then $|\mu| = \mu^+ + \mu^-$ is called the *total variation of* μ and $|\mu|(R^n) = \|\mu\|$ the *norm* of μ. (It must be noted that $|\mu|$ is a measure while $\|\mu\|$ is a real number.) The application $\mu \to \|\mu\|$ is really a norm on $\mathscr{M}_n$, that is, $\|\mu\|$ has the following properties:

(a) $\|\mu\| = 0$ if and only if $\mu = 0$;
(b) $\|\alpha\mu\| = |\alpha| \cdot \|\mu\|$ for any $\alpha \in R$;
(c) $\|\mu + \nu\| \le \|\mu\| + \|\nu\|$.

[(a) and (b) are evident while (c) follows easily from Theorem 1.4.2.]

We define $m \in R^n$ as an *increase point* of μ if $|\mu|(B) > 0$ for any Borel neighborhood of m. Then $S(\mu)$ is the set of all the increase points of μ.

---

† F_σ^n is the class of all the sets which are unions of a countable sequence of closed sets while G_δ^n is the class of all the sets which are intersections of a countable sequence of open sets.

If $\mu \in \mathscr{M}_n$ and if f is a measurable function defined on R^n with complex values, we say that f is *integrable with respect* to μ if f is integrable with respect to $|\mu|$ and we define the integral of f by

$$\int f(x)\mu(dx) = \int f(x)\mu^+(dx) - \int f(x)\mu^-(dx),$$

$\mu = \mu^+ - \mu^-$ being the Hahn–Jordan decomposition of μ. This integral is linear with respect to the function and with respect to the signed measure, but evidently is not positive. Nevertheless, we have the important inequality

$$\left| \int f(x)(dx) \right| \leq \int |f(x)|\, |\mu|(dx)$$

and also the validity of the dominated convergence theorem.

1.5 SINGULAR AND ABSOLUTELY CONTINUOUS MEASURES

Let μ and ν be two n-variate signed measures. ν is absolutely continuous with respect to μ if $\nu(B) = 0$ for every Borel set B such that $\mu(B) = 0$. Also ν is singular with respect to μ if ν is concentrated on a set A such that $|\mu|(A) = 0$. We have then

Theorem 1.5.1 (*Radon–Nikodym theorem*) If μ is an n-variate σ-finite measure and if $\nu \in \mathscr{M}_n$ is absolutely continuous with respect to μ, there exists some μ-integrable function f defined on R^n with real values such that

$$\nu(B) = \int_B f(x)\mu(dx) \tag{1.5.1}$$

for any $B \in \mathscr{B}_n$.

Proof See Halmos (1950, p. 128).

A function f satisfying (1.5.1) is the Radon–Nikodym derivative of ν with respect to μ and is denoted by $f = d\nu/d\mu$. [We must notice that this derivative is not completely determined: if $f_1 = d\nu/d\mu$, then $f_2 = d\nu/d\mu$ if and only if $\mu(\{x : f_1(x) \neq f_2(x)\}) = 0$.]

Theorem 1.5.2 (*Lebesgue decomposition theorem*) If μ is an n-variate σ-finite measure and $\nu \in \mathscr{M}_n$, then ν admits a unique decomposition

$$\nu = \nu_1 + \nu_2$$

where ν_1 is absolutely continuous and ν_2 is singular with respect to μ.

Proof See Halmos (1950, p. 134).

We define $m \in R^n$ as an *atom* of μ if

$$\mu(\{m\}) \neq 0.$$

If μ has no atom, we say that μ is *continuous*. If $\mu \neq 0$ is concentrated on an enumerable set, we say that μ is *purely atomic* (or *discrete*). We can now prove easily

Theorem 1.5.3 If $\mu \in \mathscr{M}_n$, μ admits a unique decomposition

$$\mu = \mu_1 + \mu_2$$

where μ_1 is purely atomic and μ_2 is continuous.

If we define the atomic spectrum of $\mu \in \mathscr{M}_n$ as the countable set $D(\mu)$ of all the atoms of μ, then for every $\mu \in \mathscr{M}_n$

$$S(\mu_1) = \overline{D(\mu)}$$

where μ_1 is the purely atomic measure defined in Theorem 1.5.3.

Theorem 1.5.4 If $\mu \in \mathscr{M}_n$, μ admits a unique decomposition

$$\mu = \mu_1 + \mu_2 + \mu_3$$

where μ_1 is purely atomic, μ_2 is continuous and singular, and μ_3 is absolutely continuous with respect to the Lebesgue measure.

1.6 CONTINUITY SETS

A class of sets $\mathscr{A} \subset \mathscr{B}_n$ is a determining class if

$$\mu_1(B) = \mu_2(B)$$

for any $B \in \mathscr{A}$ ($\mu_j \in \mathscr{M}_n, j = 1, 2$) implies $\mu_1 = \mu_2$.

Theorem 1.6.1 If a class of sets $\mathscr{A} \subset \mathscr{B}_n$ is closed for finite intersections and if the σ-ring generated by $\mathscr{A}$ is $\mathscr{B}_n$, then $\mathscr{A}$ is a determining class.

Proof This follows immediately from Theorem 1.1.1.

For example, the class of open (or closed) cubes with vertices having rational coordinates is an enumerable determining class.

A Borel set B is a *continuity set* of $\mu \in \mathscr{M}_n$ if

$$|\mu|(\partial B) = 0$$

where ∂B is the boundary of B. The simplest properties of these sets are given by

Theorem 1.6.2 (a) If $\mu \in \mathcal{M}_n$, the class of the continuity sets of μ is a ring.

(b) If B is a continuity set of $\mu_j \in \mathcal{M}_n$ $(j = 1, 2)$ and if α_j is a real constant $(j = 1, 2)$, then B is a continuity set of $\alpha_1 \mu_1 + \alpha_2 \mu_2$.

(c) If $\mu_j \in \mathcal{M}_n$ $(j = 1, \ldots, m)$, the class of all the open cubes which are continuity sets of μ_j $(j = 1, \ldots, m)$ is a determining class.

Proof Part (a) follows from the relations

$$\partial(B_1 \cup B_2) \subset \partial B_1 \cup \partial B_2, \qquad \partial(B_1 \cap B_2) \subset \partial B_1 \cup \partial B_2,$$

and

$$|\alpha_1 \mu_1 + \alpha_2 \mu_2| \le |\alpha_1| \cdot |\mu_1| + |\alpha_2| \cdot |\mu_2|$$

implies (b).

To prove (c), we consider first the case $m = n = 1$. Then $[a, b]$ is not a continuity set of μ if and only if one of the a and b is an atom of μ. Therefore, the set of intervals which are not continuity sets of μ is countable and, from Theorem 1.6.1, the class of open intervals which are continuity sets of μ is a determining class.

In the general case, let $C = \{x : a_j < x_j < b_j, j = 1, \ldots, n\}$. Since†

$$|\mu|(\partial C) \le \sum_{j=1}^{n} \mathrm{pr}_{e_j} |\mu| (\{a_j\} \cup \{b_j\}),$$

C is a continuity set of μ if $[a_j, b_j]$ is a continuity set of $\mathrm{pr}_{e_j}|\mu|$ $(j = 1, \ldots, n)$. Therefore C is a continuity set of μ if $a = (a_1, \ldots, a_n)$ and $b = (b_1, \ldots, b_n)$ belong to a set of the kind

$$\prod_{j=1}^{n} A_j$$

where A_j^c is countable $(j = 1, \ldots, n)$. Part (c) then follows from Theorem 1.6.1.

---

† For the definition of pr μ, see Section 2.4.

Chapter 2

Fourier–Stieltjes Transforms of Signed Measures

In this chapter, we give some fundamental results on Fourier–Stieltjes transforms of finite signed measures. These results are usually given for characteristic functions of probabilities, but the extension to Fourier–Stieltjes transforms of arbitrary finite signed measures is not difficult and often useful.

2.1 FOURIER–STIELTJES TRANSFORMS

If $\mu \in \mathscr{M}_n$, we define the *Fourier–Stieltjes* transform of μ as the function $\hat{\mu}$ defined on R^n by

$$\hat{\mu}(t) = \int e^{i(t,\, u)}\mu(du).$$

If $\mu \in \mathscr{P}_n$, its Fourier–Stieltjes transform $\hat{\mu}$ is called the *characteristic function* of μ. In the following, we do not restate that the notation $\hat{\mu}$ means the Fourier–Stieltjes transform (resp. the characteristic function) of a measure (resp. a probability) μ.

The simplest properties of this transform are given by the following theorems.

Theorem 2.1.1 The application $\mu \to \hat{\mu}$ is linear.

Theorem 2.1.2 If $\mu \in \mathscr{M}_n$, then

(a) $\hat{\mu}$ is uniformly continuous on R^n;
(b) $\hat{\mu}(0) = \mu(R^n)$;
(c) $|\hat{\mu}(t)| \leq \|\mu\|$;
(d) $\hat{\mu}(-t) = \hat{\mu}(t)$.

Proof Parts (b)–(d) are evident. To prove (a), we start from

$$|\hat{\mu}(t+h) - \hat{\mu}(t)| \leq \int |e^{i(h,u)} - 1| \, |\mu|(du) = 2\int |\sin(\tfrac{1}{2}(h,u))| \, |\mu|(du)$$
$$\leq 2\left[|\mu|(\{x : \|x\| \geq k\}) + \int_{\{\|x\| < k\}} |\sin(\tfrac{1}{2}(h,u))| \, |\mu|(du)\right].$$

For any $\varepsilon > 0$, we can choose some k such that

$$|\mu|(\{x : \|x\| \geq k\}) \leq \varepsilon/4,$$

then find some l such that

$$\int_{\{\|x\| < k\}} |\sin(\tfrac{1}{2}(h,u))| \, |\mu|(du) \leq \varepsilon/4$$

for any h satisfying $\|h\| < l$. Then

$$|\hat{\mu}(t+h) - \hat{\mu}(t)| \leq \varepsilon$$

for any $t \in R^n$, so that $\hat{\mu}$ is uniformly continuous.

We give now two inequalities for characteristic functions which are useful in the following.

Theorem 2.1.3 If $p \in \mathscr{P}_n$, then

$$\operatorname{Re}(1 - \hat{p}(2t)) \leq 4 \operatorname{Re}(1 - \hat{p}(t)).$$

Proof

$$\operatorname{Re}(1 - \hat{p}(2t)) = \int (1 - \cos(2t, u))p(du) = 2\int \sin^2(t, u)\, p(du)$$
$$\leq 4\int (1 - \cos(t, u))p(du) = 4 \operatorname{Re}(1 - \hat{p}(t)).$$

Theorem 2.1.4 If $p \in \mathscr{P}_n$, then

$$|\hat{p}(t) - \hat{p}(t+h)|^2 \leq 2(1 - \operatorname{Re} \hat{p}(h)).$$

Proof

$$|\hat{p}(t) - \hat{p}(t+h)|^2 = \left|\int e^{i(t,u)}(1 - e^{i(h,u)})p(du)\right|^2$$

and from Schwarz's inequality

$$|\hat{p}(t) - \hat{p}(t+h)|^2 \leq \int p(du) \int |1 - e^{i(h,u)}|^2 p(du)$$
$$= 2\int (1 - \cos(h, u))p(du) = 2(1 - \operatorname{Re} \hat{p}(h)).$$

Examples of characteristic functions will be given in Section 3.1.

2.2 UNIQUENESS THEOREM

Theorem 2.2.1 If $\mu_j \in \mathcal{M}_n$ $(j = 1, 2)$, the following conditions are equivalent:

(a) $\mu_1 = \mu_2$;

(b) $\int f(x)\mu_1(dx) = \int f(x)\mu_2(dx)$ for any bounded continuous function f;

(c) $\int f(x)\mu_1(dx) = \int f(x)\mu_2(dx)$ for any bounded continuous function f which is periodic with respect to each variable;

(d) $\int f(x)\mu_1(dx) = \int f(x)\mu_2(dx)$ for any bounded continuous function f having a compact support (that is, there exists a compact set K such that $f(x) = 0$ if $x \notin K$).

Proof Evidently, (a) implies (b) and (b) implies (c). We prove now that (c) implies (d). Let f be a bounded continuous function having a compact support K and, for any $m = (m_1, \ldots, m_n) \in R^n$, let f_m be the function which is periodic with period $2m_j$ with respect to the jth variable $(j = 1, \ldots, n)$ and satisfies $f_m(x) = f(x)$ for $-m < x < m$. If m is great enough, $C_m = \{x \in R^n : -m < x < m\}$ contains K and f_m is continuous. We have then, if (c) is satisfied,

$$\int f_m(x)\mu_1(dx) = \int f_m(x)\mu_2(dx).$$

From

$$\lim_{m_1\to\infty} \cdots \lim_{m_n\to\infty} \mu_j(C_m^c) = 0$$

$(j = 1, 2)$, we deduce easily that

$$\lim_{m_1\to\infty} \cdots \lim_{m_n\to\infty} \int f_m(x)\mu_j(dx) = \int f(x)\mu_j(dx)$$

$(j = 1, 2)$, so that (d) is satisfied.

Finally, we prove that (d) implies (a). Let

$$A = \{x : a \leq x \leq b\}, \qquad A_k = \left\{x : a - \frac{b-a}{k} \leq x \leq b + \frac{b-a}{k}\right\}$$

where $a \in R^n$, $b \in R^n$, $a < b$. If we consider a function f_k which is continuous, equal to 1 on A and equal to 0 on A_k^c, then

$$\lim_{k \to \infty} f_k = i_A,$$

and from the dominated convergence theorem

$$\lim_{k \to \infty} \int f_k(x)\mu_j(dx) = \int i_A(x)\mu_j(dx) = \mu_j(A)$$

($j = 1, 2$), and, if (d) is satisfied, we have

$$\mu_1(A) = \mu_2(A)$$

for any closed cube and, since the class of all the closed cubes is a determining class, $\mu_1 = \mu_2$.

Now, we prove that the application $\mu \to \hat{\mu}$ is injective. In other words, we have

Theorem 2.2.2 (*Uniqueness theorem*) If $\mu_j \in \mathcal{M}_n$ ($j = 1, 2$), then $\mu_1 = \mu_2$ if and only if $\hat{\mu}_1 = \hat{\mu}_2$.

Proof The necessity is evident. For proving the sufficiency, it follows from Theorem 2.1.1 that we must show that $\hat{\mu} = 0$ implies $\mu = 0$. By assumption, we have

$$\int e^{i(t,\, u)}\mu(du) = 0$$

for any $t \in R^n$, so that

$$\int T(u)\mu(du) = 0$$

for any trigonometric polynomial T.

Let f be a bounded continuous function which is periodic with respect to each variable. From the Stone–Weierstrass theorem (see Appendix A.1), there exists a sequence $\{T_m\}$ of trigonometric polynomials such that

$$\lim_{m \to \infty} T_m(u) = f(u),$$

this convergence being uniform. For any $\varepsilon > 0$, there exists a compact set K such that

$$\mu(K^c) < \varepsilon.$$

Then

$$\left| \int (f(u) - T_m(u))\mu(du) \right| \le \left| \int_K (f(u) - T_m(u))\mu(du) \right| + C\varepsilon$$

where

$$C = \sup_{m} \sup_{u \in R^n} | f(u) - T_m(u) |.$$

Now, from the dominated convergence theorem, there exists some m_0 such that

$$\left| \int_K (f(u) - T_m(u))\mu(du) \right| < \varepsilon$$

if $m > m_0$. This implies that

$$\int f(u)\mu(du) = \lim_{m \to \infty} \int T_m(u)\mu(du) = 0$$

for any bounded continuous function f which is periodic with respect to each variable. It follows then from the preceding theorem that $\mu = 0$.

It must be noted that the assumptions of this theorem cannot be weakened. For example, we consider the functions f_1 and f_2 defined on R by

$$f_1(t) = \begin{cases} 1 - (|t|/r) & \text{if} \quad |t| \le r, \\ 0 & \text{if} \quad |t| > r, \end{cases}$$

and

$$f_1(t) = f_2(t) \qquad \text{if} \quad |t| \le r,$$

f_2 being periodical with period $2r$. Then f_1 and f_2 are characteristic functions of univariate probabilities (for this assertion, see Theorems 3.1.1 and 3.1.2). From the uniqueness theorem, it follows that the corresponding probabilities are different. Another example of the same type will be given in Section 2.4.

If $\mu \in \mathcal{M}_n$, the signed measure $\tilde{\mu}$ defined by

$$\tilde{\mu}(B) = \mu(-B)$$

for any $B \in \mathcal{B}_n$ is the symmetric of μ. We say that $\mu \in \mathcal{M}_n$ is *symmetric* if

$$\mu = \tilde{\mu}$$

and *antisymmetric* if

$$\mu = -\tilde{\mu}.$$

We have then

Corollary 2.2.1 $\mu \in \mathcal{M}_n$ is symmetric if and only if $\hat{\mu}(t)$ is real for any $t \in R^n$ and antisymmetric if and only if $\hat{\mu}(t)$ is purely imaginary for any $t \in R^n$.

Proof The necessity of the condition is evident. To prove the sufficiency in the first case we consider the signed measure $\nu = \frac{1}{2}(\mu + \tilde{\mu})$. Evidently, ν is symmetric. If $\hat{\mu}(t) \in R$ for any $t \in R^n$, then $\hat{\mu} = \hat{\nu}$ and, from the uniqueness theorem, $\mu = \nu$. The second part can be proved by the same method.

2.3 INVERSION FORMULAS

We want to give in this section a more precise meaning to the uniqueness theorem by proving

Theorem 2.3.1 If $\mu \in \mathscr{M}_n$, $a = (a_1, \ldots, a_n) \in R^n$, $b = (b_1, \ldots, b_n) \in R^n$, $a < b$, and if $\{x \in R^n : a < x < b\}$ is a continuity set of μ, then

$$\mu(\{a < x < b\}) = \frac{1}{(2\pi)^n} \lim_{T_1 \to \infty} \cdots \lim_{T_n \to \infty} \int_{C_T} \prod_{j=1}^{n} \left(\frac{e^{-it_j a_j} - e^{-it_j b_j}}{it_j} \right) \hat{\mu}(t) \lambda(dt) \tag{2.3.1}$$

where $T = (T_1, \ldots, T_n)$, $t = (t_1, \ldots, t_n)$, $C_T = \{x \in R^n : -T \le x \le T\}$.

Proof We prove this result for measures, the general case following from this one by linearity. If we consider the function

$$f(T) = \int_{C_T} \prod_{j=1}^{n} \left(\frac{e^{-it_j a_j} - e^{-it_j b_j}}{it_j} \right) \hat{\mu}(t) \lambda(dt)$$
$$= \int_{C_T} \left(\prod_{j=1}^{n} \frac{e^{-it_j a_j} - e^{-it_j b_j}}{it_j} \right) \left(\int e^{i(t,u)} \mu(du) \right) \lambda(dt)$$

from Fubini's theorem, we can exchange the order of integrations and obtain

$$f(T) = \int \left(\int_{C_T} \prod_{j=1}^{n} \frac{e^{it_j(u_j - a_j)} - e^{it_j(u_j - b_j)}}{it_j} \lambda(dt) \right) \mu(du)$$
$$= 2^n \int \left(\prod_{j=1}^{n} \int_0^{T_j} \frac{\sin(t_j(u_j - a_j)) - \sin(t_j(u_j - b_j))}{t_j} dt_j \right) \mu(du)$$
$$= 2^n \int g(u, T) \mu(du)$$

where

$$g(u, T) = \prod_{j=1}^{n} \left(\int_0^{T_j} \frac{\sin(t_j(u_j - a_j)) - \sin(t_j(u_j - b_j))}{t_j} dt_j \right).$$

If we denote by Γ the cube $\{x \in R^n : a < x < b\}$, we have

$$\begin{aligned} f(T) &= 2^n \int_{\Gamma^c} g(u, T)\mu(du) + 2^n \int_{\partial\Gamma} g(u, T)\mu(du) + 2^n \int_{\Gamma} g(u, T)\mu(du) \\ &= I_1(T) + I_2(T) + I_3(T). \end{aligned}$$

Since Γ is a continuity set of μ, $I_2(T) = 0$.

Now, we use the following lemma [for a proof, see Lukacs (1970, p. 31)].

Lemma 2.3.1 If

$$\phi(h, l) = \int_0^l \frac{\sin(hy)}{y}\, dy,$$

then

$$\lim_{l \to \infty} \phi(h, l) = \begin{cases} \pi/2 & \text{if } h > 0, \\ 0 & \text{if } h = 0, \\ -\pi/2 & \text{if } h < 0. \end{cases}$$

If we apply this lemma to g, we obtain

$$\lim_{T_1 \to \infty} \cdots \lim_{T_n \to \infty} g(u, T) = \begin{cases} 0 & \text{if } x \notin \bar{\Gamma}, \\ \pi^n & \text{if } x \in \Gamma. \end{cases}$$

From this and the dominated convergence theorem, we obtain

$$\lim_{T_1 \to \infty} \cdots \lim_{T_n \to \infty} I_1(T) = 2^n \int_{\Gamma^c} \left(\lim_{T_1 \to \infty} \cdots \lim_{T_n \to \infty} g(u, T) \right) \mu(du) = 0,$$

$$\begin{aligned} \lim_{T_1 \to \infty} \cdots \lim_{T_n \to \infty} I_3(T) &= 2^n \int_{\Gamma} \left(\lim_{T_1 \to \infty} \cdots \lim_{T_n \to \infty} g(u, T) \right) \mu(du) \\ &= (2\pi)^n \mu(\Gamma), \end{aligned}$$

so that

$$\lim_{T_1 \to \infty} \cdots \lim_{T_n \to \infty} f(T) = (2\pi)^n \mu(\Gamma)$$

which is equivalent to (2.3.1).

From Theorems 2.3.1 and 1.6.2, we can deduce a new proof of the uniqueness theorem.

Corollary 2.3.1 If $\hat{\mu}$ is integrable, then μ is absolutely continuous with respect to the Lebesgue measure λ and its Radon–Nikodym derivative is given by

$$\frac{d\mu}{d\lambda}(x) = (1/(2\pi)^n) \int e^{-i(t,x)}\hat{\mu}(t)\lambda(dt). \tag{2.3.2}$$

Proof If $\hat{\mu}$ is integrable, we deduce from (2.3.1) that μ is absolutely continuous and letting $b \to a$ in (2.3.1), we obtain (2.3.2).

Theorem 2.3.2 If $x \in R^n$ and $\mu \in \mathscr{M}_n$, then

$$\mu(\{x\}) = \lim_{T_1\to\infty} \cdots \lim_{T_n\to\infty} \left(\prod_{j=1}^{n} (1/2T_j)\right) \int_{C_T} e^{-i(t,x)}\hat{\mu}(t)\lambda(dt) \tag{2.3.3}$$

where $T = (T_1, \ldots, T_n)$ and $C_T = \{x \in R^n : -T \le x \le T\}$.

Proof We again suppose that μ is a measure, the general case following from this one by linearity. If we define f by

$$\begin{aligned} f(T) &= \left(\prod_{j=1}^{n} (1/2T_j)\right) \int_{C_T} e^{-i(t,x)}\hat{\mu}(t)\lambda(dt) \\ &= \left(\prod_{j=1}^{n} (1/2T_j)\right) \int_{C_T} e^{-i(t,x)}\left(\int e^{i(t,u)}\mu(du)\right)\lambda(dt) \end{aligned}$$

from Fubini's theorem, we can exchange the order of integrations and obtain

$$f(T) = \int g(u, T)\mu(du)$$

where

$$g(u, T) = \prod_{j=1}^{n} (1/2T_j) \int_{-T_j}^{T_j} e^{it_j(u_j - x_j)}\, dt_j = \prod_{j=1}^{n} h(u_j, T_j)$$

with

$$h(u_j, T_j) = \begin{cases} \dfrac{\sin(T_j(u_j - x_j))}{T_j(u_j - x_j)} & \text{if } u_j \neq x_j, \\[2ex] 1 & \text{if } u_j = x_j. \end{cases}$$

Therefore

$$f(T) = \mu(\{x\}) + \int_{\{u:u\neq x\}} g(u, T)\mu(du).$$

But when $T_j \to \infty$ $(j = 1, 2, \ldots, n)$, we deduce from the dominated convergence theorem that

$$\lim_{T_1\to\infty} \cdots \lim_{T_n\to\infty} \int_{\{u:u\neq x\}} g(u, T)\mu(du) = 0$$

and this implies (2.3.3).

From this theorem and from the properties of almost periodic functions (see Appendix A.2), we have

Corollary 2.3.2 A function f defined on R^n with complex values is an almost periodic function with a summable Fourier series if and only if f is the Fourier–Stieltjes transform of a purely atomic finite signed measure.

2.4 PROJECTION THEOREM

Let E be some subspace of R^n $(E \neq \{0\})$ and $\mu \in \mathcal{M}_n$. We define the *projection of μ on E*, denoted by $\mathrm{pr}_E\, \mu$, as the signed measure defined by

$$\mathrm{pr}_E\, \mu(B) = \mu(\{x \in R^n : \mathrm{pr}_E\, x \in B\})$$

for any $B \in \mathcal{B}_n$. For convenience we define $\mathrm{pr}_{\{0\}}\, \mu$ as $\mu(R^n)\delta_0$ and when E is the one-dimensional subspace generated by $\theta \neq 0$, we write $\mathrm{pr}_\theta\, \mu$ instead of $\mathrm{pr}_E\, \mu$. By definition, $\mathrm{pr}_{e_j}\, \mu$ is the jth margin of μ. Evidently $\mathrm{pr}_E\, \mu \in \mathcal{M}_n$ and is concentrated on E. Moreover, we have

$$\mathrm{pr}_E\, \mu(R^n) = \mu(R^n). \tag{2.4.1}$$

If we choose in E a basis $\{f_1, \ldots, f_m\}$, there exists a bijection ϕ between E and R^m defined by

$$\phi\left(\sum_{j=1}^{m} \lambda_j f_j\right) = (\lambda_1, \ldots, \lambda_m) \qquad (\lambda_j \in R).$$

Then if $\nu \in \mathcal{M}(E)$, there exists an unique signed measure $\Phi(\nu) \in \mathcal{M}_m$ defined by

$$\Phi(\nu)(B') = \nu(\phi^{-1}(B'))$$

for any $B' \in \mathcal{B}_m$. Therefore any finite signed measure concentrated on E can be identified with an element of $\mathcal{M}_m$. In particular, $\mathrm{pr}_E\, \mu$ can be identified with an element of $\mathcal{M}_m$ and if E is the one-dimensional subspace generated by $\theta \neq 0$, we can identify $\mathrm{pr}_\theta\, \mu$ with a univariate finite signed measure.†

---

† Usually, $\mathrm{pr}_E\, \mu$ is defined only for one-dimensional subspaces as (with our notations) $\Phi(\mathrm{pr}_E\, \mu)$. The extension to more general projections is useful.

When we say that we can apply to $\mathrm{pr}_\theta\, \mu$ the theorems known for univariate finite signed measures, we always have this identification in mind. Now, we prove

Theorem 2.4.1 (*Projection theorem*) Let E be some subspace of R^n. If $\mu \in \mathcal{M}_n$, then

$$\widehat{\mathrm{pr}_E\, \mu} = \mathrm{pr}_E\, \hat{\mu}.$$

Proof From the definitions of projections of μ and $\hat{\mu}$ on E, we have

$$\mathrm{pr}_E\, \hat{\mu}(t) = \hat{\mu}(\mathrm{pr}_E\, t) = \int e^{i(\mathrm{pr}_E\, t,\, u)} \mu(du),$$

$$\widehat{\mathrm{pr}_E\, \mu}(t) = \int e^{i(t,\, u)}\, \mathrm{pr}_E\, \mu(du) = \int e^{i(t,\, \mathrm{pr}_E\, u)} \mu(du)$$

and since $(\mathrm{pr}_E\, t, u) = (t, \mathrm{pr}_E\, u)$, we have

$$\mathrm{pr}_E\, \hat{\mu}(t) = \widehat{\mathrm{pr}_E\, \mu}(t)$$

for any $t \in R^n$.

From the projection and uniqueness theorems we deduce easily the following corollaries.

Corollary 2.4.1 $\mu \in \mathcal{M}_n$ is concentrated on a subspace E if and only if $\hat{\mu} = \mathrm{pr}_E\, \hat{\mu}$.

Corollary 2.4.2 If E and F are two subspaces of R^n, $E \subset F$, then

$$\mathrm{pr}_F(\mathrm{pr}_E\, \mu) = \mathrm{pr}_E(\mathrm{pr}_F\, \mu) = \mathrm{pr}_{E \cap F}\, \mu.$$

Theorem 2.4.2 If m is an integer satisfying $1 \leq m < n$ and if $\mu \in \mathcal{M}_n$, μ is determined by the set of all its projections on m-dimensional subspaces of R^n.

Proof We suppose first that $m = 1$. If $\mathrm{pr}_\theta\, \mu$ is given for some $\theta \in R^n$, then, by the projection theorem, we know $\hat{\mu}(\lambda\theta)$ for any $\lambda \in R$. Therefore if $\mathrm{pr}_\theta\, \mu$ is given for any $\theta \in R^n$, we know $\hat{\mu}(t)$ for any $t \in R^n$ and, by the uniqueness theorem, μ is completely determined.

Now if $m > 1$ and if the projections of μ on m-dimensional subspaces of R^n are given, then, from Corollary 2.4.2, we can determine the projections of μ on all the one-dimensional subspaces of R^n.

When μ is a probability and $m = 1$, this result is known as the "Cramér–Wold theorem." The case $m = 1$ gives

Corollary 2.4.3 The class $\{A_{m,\theta}\}$ where

$$A_{m,\theta} = \{x \in R^n : (x, \theta) \leq m\}$$

for $\theta \in R^n$ and $m \in R$ is a determining class.

With the same proof, using the fact that a Fourier–Stieltjes transform is continuous, we obtain the following extension of Theorem 2.4.2.

Theorem 2.4.3 If $(E_l)_{l \in L}$ is a class of subspaces of R^n such that $\bigcup_{l \in L} E_l$ is dense on R^n, then the class $(\mathrm{pr}_{E_l}\, \mu)_{l \in L}$ determines the signed measure μ.

The following example shows that we cannot weaken this last condition. Let f_1 and f_2 be the functions given in Section 2.2, and let $\hat{\mu}_1$ and $\hat{\mu}_2$ be defined by

$$\hat{\mu}_1(t_1, t_2) = f_1(t_1) f_1(t_2), \qquad \hat{\mu}_2(t_1, t_2) = f_1(t_1) f_2(t_2).$$

$\hat{\mu}_1$ and $\hat{\mu}_2$ are Fourier–Stieltjes transforms of some probabilities μ_1 and μ_2 satisfying

$$\mu_1 \neq \mu_2 \qquad \text{and} \qquad \mathrm{pr}_\theta\, \mu_1 = \mathrm{pr}_\theta\, \mu_2$$

for any $\theta = (\theta_1, \theta_2)$ such that $|\theta_1 \theta_2^{-1}| \leq 1$.

The conditions of Theorem 2.4.3 can be weakened for some particular classes of measures, for example, for purely atomic measures having a finite number of atoms [see Rényi (1952), Gilbert (1955), Heppes (1956)] or for measures which are product of three measures [see Kotlarski (1971)].

2.5 CONVOLUTION THEOREM

For any $B \in \mathscr{B}_n$ and any $\mu_j \in \mathscr{M}_n$ $(j = 1, 2)$, the function

$$u \to \mu_1(B - u)$$

is a bounded function which is measurable with respect to μ_2, so that the integral

$$\int \mu_1(B - u)\mu_2(du) = \mu(B)$$

exists for any $B \in \mathscr{B}_n$. Then μ is a finite signed measure which is called the *convolution* of μ_1 and μ_2. We write then

$$\mu = \mu_1 * \mu_2 .$$

If we consider the product measure $\mu_1 \times \mu_2$ on R^{2n}, then it can be shown easily that $\mu_1 * \mu_2$ can be identified with the projection of $\mu_1 \times \mu_2$ on the diagonal subspace $\{(x, y) \in R^{2n} : x = y\}$. On the other hand, if $\mu = \mu_1 \times \mu_2$ where μ_j is an n_j-variate signed measure $(n_1 + n_2 = n)$, then $\mu = \mathrm{pr}_{E_1}\, \mu * \mathrm{pr}_{E_2}\, \mu$ where E_1 is the subspace generated by $e_1, \ldots, e_{n_1}$ and E_2 is the subspace generated by $e_{n_1+1}, \ldots, e_n$. Moreover, $\mathrm{pr}_{E_j}\, \mu$ can be identified with μ_j in the isomorphism applying E_j on R^{n_j}.

The convolution is an operation on $\mathcal{M}_n$ which is evidently linear with respect to each signed measure

$$(\alpha_1\mu_1 + \alpha'_1\mu'_1) * \mu_2 = \alpha_1(\mu_1 * \mu_2) + \alpha'_1(\mu'_1 * \mu_2),$$
$$\mu_1 * (\alpha_2\mu_2 + \alpha'_2\mu'_2) = \alpha_2(\mu_1 * \mu_2) + \alpha'_2(\mu_1 * \mu'_2)$$

for any real constants α_j and α'_j $(j = 1, 2)$. From the definition, we have also

Theorem 2.5.1 If $\mu_j \in \mathcal{M}_n$ $(j = 1, 2)$ and if $\mu = \mu_1 * \mu_2$, then

(a) If μ_j is concentrated on $A_j \in F_\sigma^n$ $(j = 1, 2)$, then μ is concentrated on $A_1 + A_2$;
(b) $\mu(R^n) = \mu_1(R^n)\mu_2(R^n)$.

Proof We first note that if $A_j \in F_\sigma^n$ $(j = 1, 2)$, then $A_1 + A_2$ belongs also to F_σ^n. Indeed, let $\{A_{j,k}\}$ $(j = 1, 2)$ be a sequence of closed bounded sets such that $\bigcup_{k=1}^{\infty} A_{j,k} = A_j$ $(j = 1, 2)$. Then $\{A_{1,k} + A_{2,k}\}$ is a sequence of closed bounded sets such that

$$\bigcup_{k=1}^{\infty} (A_{1,k} + A_{2,k}) = A_1 + A_2 .$$

Now, (a) and (b) follow immediately from the definition.

In connection with this result, we must notice that there exist some Borel sets A_j $(j = 1, 2)$ such that $A_1 + A_2$ is not Borel [for such an example, see Erdös and Stone (1970)].

Corollary 2.5.1 If μ_j is a finite measure $(j = 1, 2)$, $\mu_1 * \mu_2$ is a measure and

$$\|\mu_1 * \mu_2\| = \|\mu_1\| \cdot \|\mu_2\|.$$

In particular, if μ_j is a probability $(j = 1, 2)$, then $\mu_1 * \mu_2$ is a probability.

Corollary 2.5.2 If $\mu = \mu_1 * \mu_2$, then

$$D(\mu) \subset D(\mu_1) + D(\mu_2), \qquad S(\mu) \subset \overline{S(\mu_1) + S(\mu_2)}.$$

When μ_j is a measure $(j = 1, 2)$, the inclusions can be replaced by equalities.

For the second relation, it must be noted that the vectorial sum of two closed sets is not in general a closed set. For example, we take $A_1 = N$, $A_2 = \{x \in R : x = -n + (1/n), n \in N, n \geq 2\}$. Then

$$\{x \in R^n : x = 1/n, n \in N, n \geq 2\} \subset A_1 + A_2 ,$$

so that $0 \in \overline{(A_1 + A_2)} \backslash (A_1 + A_2)$. Nevertheless, if the two closed sets are bounded, then their vectorial sum is closed.

From Corollary 2.5.2, we deduce immediately the following result which will be useful.

Corollary 2.5.3 If p_j is a probability ($j = 1, 2$) and if α is an increase point of p_2, then

$$S(p_1) \subset S(p_1 * p_2) - \alpha.$$

If α is an atom of p_2, then

$$D(p_1) \subset D(p_1 * p_2) - \alpha.$$

Corollary 2.5.4 If $\mu_j \in \mathcal{M}_n$ ($j = 1, 2$) and if $\mu = \mu_1 * \mu_2$, then

$$\|\mu\| \leq \|\mu_1\| \cdot \|\mu_2\|.$$

Proof If $\mu = \mu^+ - \mu^-$ and $\mu_j = \mu_j^+ - \mu_j^-$ ($j = 1, 2$) are the Hahn–Jordan decompositions of μ and μ_j ($j = 1, 2$), respectively, then, from Theorem 1.4.2,

$$\mu^+ \leq (\mu_1^+ * \mu_2^+) + (\mu_1^- * \mu_2^-), \qquad \mu^- \leq (\mu_1^+ * \mu_2^-) + (\mu_1^- * \mu_2^+),$$

and from these inequalities and Corollary 2.5.1, we deduce the corollary.

From the definition, we have also

Theorem 2.5.2 If $\mu_1 \in \mathcal{M}_n$ is absolutely continuous with respect to the Lebesgue measure and if $\mu_2 \in \mathcal{M}_n$, then $\mu = \mu_1 * \mu_2$ is absolutely continuous and

$$\frac{d\mu}{d\lambda}(x) = \int \frac{d\mu_1}{d\lambda}(x - u)\mu_2(du).$$

Corollary 2.5.5 If $\mu_j \in \mathcal{M}_n$ is absolutely continuous with respect to the Lebesgue measure ($j = 1, 2$) and if $\mu = \mu_1 * \mu_2$, then

$$\frac{d\mu}{d\lambda}(x) = \int \frac{d\mu_1}{d\lambda}(x - u)\frac{d\mu_2}{d\lambda}(u)\lambda(du).$$

From Fubini's theorem, we deduce easily

Theorem 2.5.3 Let f be a function defined on R^n with complex values and let μ_j be a finite measure ($j = 1, 2$). If f is integrable with respect to $\mu = \mu_1 * \mu_2$, then

$$\int f(u)\mu(du) = \int \left[\int f(v + w)\mu_1(dv)\right]\mu_2(dw)$$
$$= \int \left[\int f(v + w)\mu_2(dw)\right]\mu_1(dv),$$

this relation meaning that the integrals within brackets exist almost everywhere and are integrable.

Corollary 2.5.6 If $\mu = \mu_1 * \mu_2$, then

$$\mu(B) = \int_{\{v+w \in B\}} \mu_1(dv)\mu_2(dw)$$

for any $B \in \mathscr{B}_n$.

Proof When μ_j is a measure ($j = 1, 2$), we obtain the corollary by letting, in Theorem 2.5.3, $f = i_B$. The general case follows from this one by linearity.

This corollary can also be deduced from the fact that $\mu_1 * \mu_2$ can be considered as the projection of $\mu_1 \times \mu_2$ on the diagonal subspace of R^{2n}.

Now we prove the important

Theorem 2.5.4 (*Convolution theorem*) If $\mu_j \in \mathscr{M}_n$ ($j = 1, 2$), then $\mu = \mu_1 * \mu_2$ if and only if $\hat{\mu} = \hat{\mu}_1 \cdot \hat{\mu}_2$.

Proof If μ_1 and μ_2 are finite measures and if $\mu = \mu_1 * \mu_2$, applying Theorem 2.5.3 to $f_t(u) = e^{i(t,u)}$, we obtain

$$\hat{\mu}(t) = \int e^{i(t,u)}\mu(du) = \int \left[\int e^{i(t,v+w)}\mu_1(dv)\right]\mu_2(dw)$$

$$= \int e^{i(t,v)}\mu_1(dv) \cdot \int e^{i(t,w)}\mu_2(dw) = \hat{\mu}_1(t)\hat{\mu}_2(t)$$

for any $t \in R^n$, so that $\hat{\mu} = \hat{\mu}_1 \hat{\mu}_2$. From this we deduce the general case by linearity. The converse follows immediately from the uniqueness theorem.

Corollary 2.5.7 The product of the Fourier–Stieltjes transforms of two signed measures is the Fourier–Stieltjes transform of a signed measure. The product of the characteristic functions of two probabilities is the characteristic function of a probability.

From the projection and convolution theorems, we have

Corollary 2.5.8 If $\mu_j \in \mathscr{M}_n$ ($j = 1, 2$) and if E is a subspace of R^n, then

$$\text{pr}_E(\mu_1 * \mu_2) = \text{pr}_E\, \mu_1 * \text{pr}_E\, \mu_2 .$$

From convolution theorem, we deduce also that the convolution is associative and commutative and has unit δ_0. More precisely, we have from Theorems 2.5.1 and 2.5.4

Theorem 2.5.5 If A is a Borel set of R^n, $\mathscr{M}((N)A)$ is a commutative ring for addition and convolution.

Now, we give a more specialized result by introducing the following definition.

If $m = (m_1, \ldots, m_n) \in R^n$, $\mu \in \mathscr{M}_n$ is a lattice signed measure with span m if there exists some $\alpha \in R^n$ such that μ is concentrated on $(Z)A + \alpha$ where $A = \{m_1 e_1, \ldots, m_n e_n\}$.

We can characterize this class of signed measures by

Theorem 2.5.6 $\mu \in \mathcal{M}_n$ is a lattice signed measure with span $m = (m_1, \ldots, m_n)$ if and only if there exists some $\alpha \in R^n$ such that $\hat{\mu} e^{-i(\alpha, \cdot)}$ is a function which is periodic with period k_j with respect to the jth variable where $k_j = 2\pi/m_j$ if $m_j \neq 0$ and k_j is arbitrary if $m_j = 0$ $(j = 1, \ldots, n)$.

Proof The necessity is evident. We prove now the sufficiency. If $\mu * \delta_{-\alpha} = \mu_\alpha$, $\hat{\mu}_\alpha$ is periodic with period k_j with respect to the jth variable $(j = 1, \ldots, n)$. We have then

$$\hat{\mu}_\alpha(t + \tfrac{1}{2}k_j e_j) - \hat{\mu}_\alpha(t - \tfrac{1}{2}k_j e_j) = 2i \int e^{i(t, u)} \sin(\tfrac{1}{2}k_j e_j, u)\mu_\alpha(du) = 0.$$

From the uniqueness theorem, we obtain

$$\int_B \sin(\tfrac{1}{2}k_j u_j)\mu_\alpha(du) = 0$$

for any $B \in \mathcal{B}_n$, so that μ_α is concentrated on

$$A_j = \{x \in R^n : \sin(\tfrac{1}{2}k_j x_j) = 0\} = \{x \in R^n : k_j x_j = 2l\pi, l \in Z\}.$$

Therefore μ_α is concentrated on $\bigcap_{j=1}^n A_j = (Z)A$ and this, with Corollary 2.5.2, implies the theorem.

In other words, the class of the Fourier–Stieltjes transforms of lattice signed measures is a slight generalization of the class of n-variate absolutely convergent trigonometric series. For probabilities, the condition of the preceding theorem can be simplified.

Theorem 2.5.7 $p \in \mathcal{P}_n$ is a lattice probability with span $m = (m_1, \ldots, m_n) \in R^n$ if and only if

$$|\hat{p}(k_j e_j)| = 1 \qquad (j = 1, \ldots, n)$$

where $k_j = 2\pi/m_j$ if $m_j \neq 0$ and k_j is arbitrary if $m_j = 0$ $(j = 1, \ldots, n)$.

Proof For the case $n = 1$, see Lukacs (1970, Theorem 2.1.4). From this case, it follows easily that the jth margin of p is concentrated on $\{x = (x_1, \ldots, x_n) \in R^n : x_j = \alpha_j + lm_j, l \in Z\}$ and, since p is a probability, we easily obtain the theorem.

2.6 CONTINUITY THEOREMS

Let $\mu_j \in \mathcal{M}_n$ $(j = 1, 2, \ldots)$ and $\mu \in \mathcal{M}_n$. We say that the sequence $\{\mu_j\}$ *converges weakly to* μ if

$$\lim_{j \to \infty} \mu_j(B) = \mu(B)$$

for any bounded continuity set B of μ and we write then

$$\mu_j \xrightarrow{w} \mu \qquad \text{or} \qquad \lim_{j\to\infty} \mu_j \overset{w}{=} \mu.$$

We say that the sequence $\{\mu_j\}$ *converges completely to* μ if $\{\mu_j\}$ converges weakly to μ and

$$\lim_{j\to\infty} \|\mu_j\| = \|\mu\|.$$

We write then

$$\mu_j \xrightarrow{c} \mu \qquad \text{or} \qquad \lim_{j\to\infty} \mu_j \overset{c}{=} \mu.$$

If $\mu_j \in \mathscr{M}_n$ ($j = 1, 2, \ldots$), the sequence $\{\mu_j\}$ is uniformly bounded if there exists some constant M such that

$$\|\mu_j\| \le M \qquad (j = 1, 2, \ldots).$$

The sequence $\{\mu_j\}$ is tight if for any $\varepsilon > 0$, there exists some compact set K such that

$$|\mu_j|(K^c) \le \varepsilon \qquad (j = 1, 2, \ldots).$$

Lemma 2.6.1 If $\mu_j \xrightarrow{c} \mu$, then $\{\mu_j\}$ is uniformly bounded.

Proof If $\mu_j \xrightarrow{c} \mu$,

$$\lim_{j\to\infty} \|\mu_j\| = \|\mu\|,$$

so that $\{\|\mu_j\|\}$ is a bounded set of real numbers.

Lemma 2.6.2 Let μ_j ($j = 1, 2, \ldots$) and μ be some finite measures and $\mathscr{A} \subset \mathscr{B}_n$ be a class of bounded open sets satisfying the following conditions:

(a) $\mathscr{A}$ is closed for finite intersections;
(b) Any bounded open set is an enumerable union of elements of $\mathscr{A}$;
(c) For any bounded closed set V, there exist $A_1, \ldots, A_m$ belonging to $\mathscr{A}$ such that

$$V \subset \bigcup_{k=1}^{m} A_k.$$

If

$$\lim_{j\to\infty} \mu_j(A) = \mu(A),$$

for any $A \in \mathscr{A}$, then $\mu_j \xrightarrow{w} \mu$.

Proof If $A_1, \ldots, A_m$ belong to $\mathscr{A}$, then

$$\lim_{j\to\infty} \mu_j\left(\bigcup_{k=1}^{m} A_k\right) = \mu\left(\bigcup_{k=1}^{m} A_k\right).$$

Indeed

$$\begin{aligned}
\mu_j\left(\bigcup_{k=1}^{m} A_k\right) &= \sum_{1\le k\le m} \mu_j(A_k) - \sum_{1\le k_1\le k_2\le m} \mu_j(A_{k_1} \cap A_{k_2}) \\
&\quad + \sum_{1\le k_1\le k_2\le k_3\le m} \mu_j(A_{k_1} \cap A_{k_2} \cap A_{k_3}) - \cdots \\
&\to \sum_{1\le k\le m} \mu(A_k) - \sum_{1\le k_1\le k_2\le m} \mu(A_{k_1} \cap A_{k_2}) \\
&\quad + \sum_{1\le k_1\le k_2\le k_3\le m} \mu(A_{k_1} \cap A_{k_2} \cap A_{k_3}) - \cdots \\
&= \mu\left(\bigcup_{k=1}^{m} A_k\right).
\end{aligned}$$

If U is a bounded open set, then

$$U = \bigcup_{k=1}^{\infty} A_k,$$

where $A_k \in \mathscr{A}$ $(k = 1, 2, \ldots)$. For any $\varepsilon > 0$, there exists some m such that

$$\mu\left(\bigcup_{k=1}^{m} A_k\right) \ge \mu(U) - \varepsilon.$$

Consequently

$$\mu(U) - \varepsilon \le \mu\left(\bigcup_{k=1}^{m} A_k\right) = \lim_{j\to\infty} \mu_j\left(\bigcup_{k=1}^{m} A_k\right) \le \liminf_{j\to\infty} \mu_j(U)$$

so that

$$\mu(U) \le \liminf_{j\to\infty} \mu_j(U)$$

for any bounded open set U.

If V is a bounded closed set, there exists $A_1, \ldots, A_m$ belonging to $\mathscr{A}$ such that

$$V \subset \bigcup_{k=1}^{m} A_k = A.$$

We have

$$\lim_{j\to\infty} \mu_j(A) = \mu(A)$$

and since A is a bounded open set

$$\mu(A \cap V^c) \leq \liminf_{j\to\infty} \mu_j(A \cap V^c).$$

From this and from

$$V = A \backslash (A \cap V^c),$$

we obtain

$$\limsup_{j\to\infty} \mu_j(V) \leq \mu(V)$$

for any bounded closed set V.

Let now A be a bounded continuity set of μ. Then

$$\mu(\mathring{A}) \leq \liminf_{j\to\infty} \mu_j(\mathring{A}) \leq \liminf_{j\to\infty} \mu_j(A)$$
$$\leq \limsup_{j\to\infty} \mu_j(A) \leq \limsup_{j\to\infty} \mu_j(\bar{A}) \leq \mu(\bar{A})$$

implies

$$\lim_{j\to\infty} \mu_j(A) = \mu(A)$$

and $\{\mu_j\}$ converges weakly to μ.

Lemma 2.6.3 If $\{\mu_j\}$ is an uniformly bounded sequence of measures and if $\{\mu_j\}$ converges weakly to μ, then

$$\mu(B) \leq \liminf_{j\to\infty} \mu_j(B)$$

for any continuity set B of μ. In particular

$$\|\mu\| \leq \liminf_{j\to\infty} \|\mu_j\|.$$

Proof Let $\{B_k\}$ be a sequence of bounded continuity sets of μ such that

$$B_k \subset B_{k+1}, \qquad B = \bigcup_{k=1}^{\infty} B_k.$$

Since B_k is a continuity set of μ,

$$\mu(B_k) = \lim_{j\to\infty} \mu_j(B_k)$$

and since μ_j is a measure ($j = 1, 2, \ldots$),

$$\mu_j(B_k) \leq \mu_j(B).$$

From these relations, we deduce that

$$\mu(B_k) \le \liminf_{j\to\infty} \mu_j(B)$$

for any k. Since

$$\mu(B) = \lim_{k\to\infty} \mu(B_k),$$

we obtain the first assertion. The second one follows from the fact that R^n is a continuity set of μ.

Lemma 2.6.4 If $\{\mu_j\}$ is a sequence of measures which is tight and converges weakly to μ, then $\{\mu_j\}$ converges completely to μ.

Proof For any $\varepsilon > 0$, there exists some compact set K such that $\mu_j(K^c) \le \varepsilon$, so that

$$\limsup_{j\to\infty} \mu_j(K^c) \le \varepsilon.$$

We can choose K as a continuity set of μ, so that

$$\limsup_{j\to\infty} \|\mu_j\| = \lim_{j\to\infty} \mu_j(K) + \limsup_{j\to\infty} \mu_j(K^c)$$
$$\le \mu(K) + \varepsilon \le \|\mu\| + \varepsilon.$$

Since ε is arbitrary, we have

$$\limsup_{j\to\infty} \|\mu_j\| \le \|\mu\|$$

and from Lemma 2.6.3

$$\liminf_{j\to\infty} \|\mu_j\| \ge \|\mu\|$$

so that $\{\mu_j\}$ converges completely to μ.

Theorem 2.6.1 (*Helly's first theorem*) If $\{\mu_j\}$ is a uniformly bounded sequence of measures, there exists a subsequence of $\{\mu_j\}$ which converges weakly to some measure μ. If $\{\mu_j\}$ is a sequence of measures which is uniformly bounded and tight, there exists a subsequence of $\{\mu_j\}$ which converges completely to some measure μ.

Proof We prove the first assertion, the second following immediately from this one and Lemma 2.6.4. Let F_j be the distribution function of μ_j $(j = 1, 2, \ldots)$ and $M = \{m_1, m_2, \ldots\}$ be an enumerable set of points of R^n such that the class $\mathscr{A}$ of the sets $\{x : m < x < m'\}$ for any $m \in M$, $m' \in M$, $m < m'$ satisfies the conditions of Lemma 2.6.2 (for example, M can be the set of all the points with rational coordinates).

Since $\{F_j(m_1)\}$ is a bounded set of real numbers, there exists a subsequence $\{F_{1,j}(m_1)\}$ of $\{F_j(m_1)\}$ which converges to, say, $F(m_1)$. From $\{F_{1,j}(m_2)\}$, we can extract a subsequence $\{F_{2,j}(m_2)\}$ which converges to $F(m_2)$ and so on. At the kth step $\{F_{k-1,j}(m_k)\}$ contains a subsequence $\{F_{k,j}(m_k)\}$ which converges to $F(m_k)$. If we consider the sequence $\{F_{j,j}\}$, then

$$\lim_{j\to\infty} F_{j,j}(m_k) = F(m_k) \tag{2.6.1}$$

for any k. If we define $\bar{F}(x)$ for any $x \in R^n$ by

$$\bar{F}(x) = \sup_{\xi \le x,\, \xi \in M} F(\xi),$$

$\bar{F}$ is an n-variate distribution function and we let μ be the corresponding measure. From (2.6.1) we deduce that

$$\lim_{j\to\infty} \mu_{j,j}(A) = \mu(A)$$

for any $A \in \mathscr{A}$ and, from Lemma 2.6.2, this implies the weak convergence of $\{\mu_{j,j}\}$ to μ.

This result is not true for arbitrary signed measures: an example is given by the sequence

$$\mu_j = \delta_0 - \delta_{\alpha/j}$$

where $\alpha \in R^n\backslash\{0\}$, which does not tend to 0 since

$$\mu_j(\{0\}) = 1.$$

Corollary 2.6.1 If $\{\mu_j\}$ is an uniformly bounded sequence of measures and if

$$\lim_{k\to\infty} \mu_{j_k} \overset{w}{=} \mu$$

for any weakly convergent subsequence $\{\mu_{j_k}\}$ of $\{\mu_j\}$, then

$$\lim_{j\to\infty} \mu_j \overset{w}{=} \mu.$$

If $\{\mu_j\}$ is an uniformly bounded and tight sequence of measures and if

$$\lim_{k\to\infty} \mu_{j_k} \overset{c}{=} \mu$$

for any completely convergent subsequence $\{\mu_{j_k}\}$ of $\{\mu_j\}$, then

$$\lim_{j\to\infty} \mu_j \overset{c}{=} \mu.$$

Proof We prove the first assertion. If $\{\mu_j\}$ does not tend to μ, there exists some bounded continuity set A of μ and some $\varepsilon > 0$ such that for any integer

K there exists some $j_k > K$ such that

$$|\mu_{j_k}(A) - \mu(A)| > \varepsilon.$$

From Theorem 2.6.1, we can extract from $\{\mu_{j_k}\}$ a subsequence $\{\mu_{j_{k'}}\}$ which converges weakly to some measure and this measure is not μ since $\{\mu_{j_{k'}}(A)\}$ does not converge to $\mu(A)$. The contradiction proves the first assertion of the corollary. The second assertion can be proved with the same method.

Theorem 2.6.2 Let $\mu_j = \mu_j^+ - \mu_j^-$ $(j = 1, 2, \ldots)$ and $\mu = \mu^+ - \mu^-$ be the Hahn–Jordan decompositions of $\mu_j \in \mathcal{M}_n$ and $\mu \in \mathcal{M}_n$, respectively. If

$$\mu_j \xrightarrow{c} \mu,$$

then

$$\mu_j^+ \xrightarrow{c} \mu^+, \qquad \mu_j^- \xrightarrow{c} \mu^-, \qquad |\mu_j| \xrightarrow{c} |\mu|.$$

Proof If $\mu_j \xrightarrow{c} \mu$, from Lemma 2.6.1, $\{\mu_j\}$ is an uniformly bounded sequence, so that $\{\mu_j^+\}$ and $\{\mu_j^-\}$ are uniformly bounded sequence of measures. Since $\{\mu_j^+\}$ is uniformly bounded, from Helly's first theorem, we can find a subsequence $\{\mu_{1,j}^+\}$ of $\{\mu_j^+\}$ which converges weakly to some measure ν_1 and since $\{\mu_{1,j}^-\}$ is uniformly bounded, we can find a subsequence $\{\mu_{2,j}^-\}$ of $\{\mu_{1,j}^-\}$ which converges weakly to some measure ν_2.

If B is a continuity set of ν_1 and ν_2, then

$$\nu_1(B) - \nu_2(B) = \mu(B)$$

and, since the class of bounded continuity sets of ν_1 and ν_2 is a determining class,

$$\nu_1 - \nu_2 = \mu.$$

From Theorem 1.4.2, we have

$$\mu^+ \leq \nu_1, \qquad \mu^- \leq \nu_2,$$

so that

$$\mu^+(B) + \mu^-(B) \leq \nu_1(B) + \nu_2(B) \tag{2.6.2}$$

for any Borel set B. From (2.6.2), we have

$$\|\mu\| = \|\mu^+ + \mu^-\| \leq \|\nu_1 + \nu_2\| \tag{2.6.3}$$

and if, for some B, the inequality in (2.6.2) is strict, then the inequality in (2.6.3) is strict also.

From the relations

$$\mu_{2,j}^+ \xrightarrow{w} \nu_1, \qquad \mu_{2,j}^- \xrightarrow{w} \nu_2 \tag{2.6.4}$$

and from Lemma 2.6.2, we have

$$\mu_{2,j}^{+} + \mu_{2,j}^{-} \xrightarrow{w} \nu_1 + \nu_2$$

and from Lemma 2.6.3

$$\|\nu_1 + \nu_2\| \leq \liminf_{j\to\infty} \|\mu_{2,j}\|. \tag{2.6.5}$$

Since $\mu_j \xrightarrow{c} \mu$ implies

$$\lim_{j\to\infty} \|\mu_j\| = \|\mu\|,$$

we obtain from (2.6.3) and (2.6.5)

$$\|\mu^+ + \mu^-\| = \|\nu_1 + \nu_2\|,$$

so that $\mu^+ = \nu_1$, $\mu^- = \nu_2$. Since, in relations (2.6.4), the limits are independent of the subsequence, it follows from Corollary 2.6.1 that

$$\mu_j^+ \xrightarrow{c} \mu^+, \qquad \mu_j^- \xrightarrow{c} \mu^-, \qquad \text{and} \qquad |\mu_j| = \mu_j^+ + \mu_j^- \xrightarrow{c} |\mu|.$$

This result is not true for weak convergence. For example, we consider the sequence

$$\mu_j = \delta_{\alpha/j} - \delta_{-\alpha/j}$$

where $\alpha \in R^n \setminus \{0\}$. Then

$$\mu_j \xrightarrow{w} 0,$$

but

$$\mu_j^+ \xrightarrow{w} \delta_0, \qquad \mu_j^- \xrightarrow{w} \delta_0, \qquad |\mu_j| \xrightarrow{w} 2\delta_0.$$

(This example proves also that Lemma 2.6.4 is not true for arbitrary signed measures.)

Theorem 2.6.3 Let $\mu_j \in \mathscr{M}_n$ $(j = 1, 2, \ldots)$ and $\mu \in \mathscr{M}_n$. If $\{\mu_j\}$ converges completely to μ, then

$$\lim_{j\to\infty} \mu_j(B) = \mu(B) \tag{2.6.6}$$

for any continuity set B of μ. Conversely, if $\{\mu_j\}$ is a sequence of measures satisfying (2.6.6) for any continuity set B of μ, then $\{\mu_j\}$ converges completely to μ.

Proof From Theorem 2.6.2, it suffices to prove this result when μ_j and μ are measures. If $\mu_j \xrightarrow{c} \mu$, we have from Lemma 2.6.3,

$$\mu(B) \leq \liminf_{j\to\infty} \mu_j(B) \qquad \text{and} \qquad \mu(B^c) \leq \liminf_{j\to\infty} \mu_j(B^c)$$

for any continuity set B of μ. If $\mu_j \xrightarrow{c} \mu$, we obtain

$$\limsup_{j\to\infty} \mu_j(B) \leq \mu(B) \leq \liminf_{j\to\infty} \mu_j(B)$$

so that

$$\mu(B) = \lim_{j\to\infty} \mu_j(B)$$

for any continuity set B of μ. The second part is evident.

The last example proves that this second result is not true for arbitrary signed measures.

Corollary 2.6.2 If $\mu_j \xrightarrow{c} \mu$, then $\{\mu_j\}$ is tight.

Proof For any $\varepsilon > 0$, there exists some compact continuity set K_0 of μ such that

$$|\mu|(K_0^c) \leq \varepsilon/2.$$

If $\mu_j \xrightarrow{c} \mu$, then from theorems 2.6.2 and 2.6.3

$$\lim_{j\to\infty} |\mu_j|(K_0^c) = |\mu|(K_0^c)$$

and there exists some j_0 such that

$$\big|\, |\mu_j|(K_0^c) - |\mu|(K_0^c) \big| \leq \varepsilon/2$$

and therefore

$$|\mu_j|(K_0^c) \leq \varepsilon$$

for any $j > j_0$. Now we can find a compact set K_j such that

$$|\mu_j|(K_j^c) \leq \varepsilon \qquad (j = 1, \ldots, j_0).$$

If we define K by

$$K = \bigcup_{j=0}^{j_0} K_j,$$

we have

$$|\mu_j|(K^c) \leq \varepsilon$$

for any j and $\{\mu_j\}$ is tight.

Corollary 2.6.3 If $\{\mu_j\}$ is a sequence of finite signed measures which converges completely to μ and if E is a subspace of R^n, then

$$\mathrm{pr}_E\, \mu_j \xrightarrow{c} \mathrm{pr}_E\, \mu.$$

Proof We suppose first that μ_j are measures. $B \in \mathscr{B}_n$ is a continuity set of $\text{pr}_E\,\mu$ if and only if $\{x \in R^n : \text{pr}_E\, x \in B\}$ is a continuity set of μ. If $\mu_j \xrightarrow{c} \mu$, we have from Theorem 2.6.4

$$\lim_{j\to\infty} \mu_j(\{x : \text{pr}_E\; x \in B\}) = \mu(\{x : \text{pr}_E\; x \in B\})$$

which is equivalent to

$$\lim_{j\to\infty} \text{pr}_E\; \mu_j(B) = \text{pr}_E\; \mu(B)$$

for any continuity set B of $\text{pr}_E\;\mu$ and from the second part of Theorem 2.6.4, this implies

$$\text{pr}_E\; \mu_j \xrightarrow{c} \text{pr}_E\; \mu.$$

The general case follows from this one and Theorem 2.6.2.

This result is not true for weak convergence. If we consider the sequence

$$\mu_j = \delta_{j\alpha}$$

for $\alpha \in R^n \backslash \{0\}$, then

$$\mu_j \xrightarrow{w} 0,$$

but

$$\text{pr}_E\; \mu_j = \delta_0$$

for any subspace E orthogonal to α.

Theorem 2.6.4 (*Helly's second theorem*) Let $\mu_j \in \mathscr{M}_n$ $(j = 1, 2, \ldots)$. If $\{\mu_j\}$ is an uniformly bounded sequence which converges weakly to μ and if f is a bounded continuous function defined on R^n with complex values, then

$$\lim_{j\to\infty} \int_B f(u)\mu_j(du) = \int_B f(u)\mu(du)$$

for any bounded continuity set B of μ.

Proof For any $\varepsilon > 0$, there exist disjoint sets $B_1, \ldots, B_m$ such that

$$B = \bigcup_{k=1}^{m} B_k$$

and

$$|f(u) - f(u')| \leq \varepsilon$$

for any $u \in B_k$, $u' \in B_k$ $(k = 1, \ldots, m)$. Since the intersection of two continuity sets is a continuity set, we can suppose that B_k is a continuity set of μ $(k = 1, \ldots, m)$. If we choose $x_k \in B_k$ and define ϕ by

$$\phi = \sum_{k=1}^{m} f(x_k) i_{B_k},$$

we have

$$\begin{aligned} \left| \int_B f(u)\mu_j(du) - \int_B \phi(u)\mu_j(du) \right| &\le \varepsilon \, |\mu_j|(B), \\ \left| \int_B f(u)\mu(du) - \int_B \phi(u)\mu(du) \right| &\le \varepsilon \, |\mu|(B). \end{aligned} \tag{2.6.7}$$

Now

$$\int_B \phi(u)\mu_j(du) - \int_B \phi(u)\mu(du) = \sum_{k=1}^{m} f(x_k)(\mu_j(B_k) - \mu(B_k)),$$

so that

$$\left| \int_B \phi(u)\mu_j(du) - \int_B \phi(u)\mu(du) \right| \le \sup_{u \in B} f(u) \sum_{k=1}^{m} |\mu_j(B_k) - \mu(B_k)|.$$

There exists some j_0 such that

$$|\mu_j(B_k) - \mu(B_k)| \le \varepsilon/m \qquad (k = 1, \ldots, m)$$

for any $j > j_0$. Then

$$\left| \int_B \phi(u)\mu_j(du) - \int_B \phi(u)\mu(du) \right| \le \varepsilon \sup_{u \in B} |f(u)|$$

and this, with (2.6.7), implies

$$\left| \int_B f(u)\mu_j(du) - \int_B f(u)\mu(du) \right| \le C\varepsilon$$

for any $j > j_0$ where

$$C = |\mu|(B) + \sup_j \|\mu_j\| + \sup_{u \in B} |f(u)|$$

and the theorem is proved.

Theorem 2.6.5 Let $\mu_j \in \mathscr{M}_n$ $(j = 1, 2, \ldots)$. If $\{\mu_j\}$ is an uniformly bounded sequence which converges weakly to μ, then

$$\lim_{j \to \infty} \int f(u)\mu_j(du) = \int f(u)\mu(du) \tag{2.6.8}$$

for any bounded continuous function f defined on R^n with complex values having a compact support K. Conversely, if $\{\mu_j\}$ is a sequence of measures such that (2.6.8) holds for any bounded continuous function f defined on R^n with complex values having a compact support, then $\{\mu_j\}$ converges weakly to μ.

Proof The first part follows from Theorem 2.6.5 since we can choose the compact set K as a continuity set of μ.

For proving the converse, we consider a cube $A = \{x \in R^n : a < x < b\}$ ($a \in R^n$, $b \in R^n$) and define A_k and A'_k by

$$A_k = \{x : a(1 + \varepsilon_k) \leq x \leq b(1 - \varepsilon_k)\},$$
$$A'_k = \{x : a(1 - \varepsilon_k) \leq x \leq b(1 + \varepsilon_k)\}$$

where $\{\varepsilon_k\}$ is a sequence of positive constants such that

$$\lim_{k \to \infty} \varepsilon_k = 0$$

and A_k and A'_k are continuity sets of μ ($k = 1, 2, \ldots$). If we consider a continuous function f_k equal to 1 on A_k and to 0 on A^c and a continuous function f'_k equal to 1 on $\bar{A}$ and to 0 on $(A'_k)^c$, we obtain

$$\int f_k(u)\mu_j(du) \leq \mu_j(A) \leq \int f'_k(u)\mu_j(du).$$

Letting $j \to \infty$, we obtain from the hypothesis

$$\int f_k(u)\mu(du) \leq \liminf_{j \to \infty} \mu_j(A) \leq \limsup_{j \to \infty} \mu_j(A) \leq \int f'_k(u)\mu(du)$$

and letting $k \to \infty$, we obtain from the dominated convergence theorem

$$\lim_{k \to \infty} \int f_k(u)\mu(du) = \mu(A), \qquad \lim_{k \to \infty} \int f'_k(u)\mu(du) = \mu(\bar{A})$$

and, when A is a continuity set of μ, this implies

$$\lim_{j \to \infty} \mu_j(A) = \mu(A)$$

and, from Lemma 2.6.2, $\{\mu_j\}$ converges weakly to μ.

The example following Theorem 2.6.1 proves that the second part of this theorem is not true for arbitrary signed measures.

We now extend this result to complete convergence.

Theorem 2.6.6 Let $\mu_j \in \mathscr{M}_n$ ($j = 1, 2, \ldots$). If $\{\mu_j\}$ converges completely to μ, then (2.6.8) holds for any bounded continuous function f defined on R^n with complex values. Conversely, if $\{\mu_j\}$ is a sequence of measures such that (2.6.8) holds for any bounded continuous function f defined on R^n with complex values, then $\{\mu_j\}$ converges completely to μ.

Proof We suppose that $\{\mu_j\}$ converges completely to μ. From Corollary 2.6.2, $\{\mu_j\}$ is tight: for any $\varepsilon > 0$, there exists some bounded continuity set B of μ such that

$$|\mu_j|(B^c) \leq \varepsilon \qquad \text{and} \qquad |\mu|(B^c) \leq \varepsilon.$$

We have then

$$\left| \int_{B^c} f(u)\mu_j(du) - \int_{B^c} f(u)\mu(du) \right| \leq 2\varepsilon \sup_{u \in R^n} |f(u)|.$$

Now, from Theorem 2.6.4, there exists some j_0 such that

$$\left| \int_B f(u)\mu_j(du) - \int_B f(u)\mu(du) \right| \leq \varepsilon$$

for $j > j_0$ and the first part of the theorem follows immediately from these two inequalities.

For proving the converse, we deduce from Theorem 2.5.5 that $\{\mu_j\}$ converges weakly to μ and applying (2.6.8) to the constant function equal to 1, we obtain

$$\lim_{j\to\infty} \|\mu_j\| = \|\mu\|.$$

The example following Theorem 2.6.1 proves that the second part is not true for arbitrary signed measures. We must also note that the assumptions of this theorem cannot be weakened. For example, if we consider the sequence

$$\mu_j = \delta_{j\alpha} - \delta_{(j+1)\alpha}$$

($\alpha \in R^n \backslash \{0\}$), then $\mu_j \xrightarrow{w} 0$ and $\mu_j(R^n) \to 0$. Nevertheless

$$\hat{\mu}_j(t) = e^{i(j\alpha,\, t)}(1 - e^{i(\alpha,\, t)})$$

does not tend in general to 0.

Applying this theorem to $f_t(u) = e^{i(t,\, u)}$ $(t \in R^n)$, we deduce that $\mu_j \xrightarrow{c} \mu$ implies $\hat{\mu}_j \to \hat{\mu}$, but we can prove more with

Theorem 2.6.7 Let $\mu_j \in \mathscr{M}_n$ $(j = 1, 2, \ldots)$. If $\{\mu_j\}$ converges completely to μ, then

$$\lim_{j\to\infty} \hat{\mu}_j(t) = \hat{\mu}(t)$$

uniformly for t belonging to a compact set of R^n.

Proof If $\varepsilon > 0$ is an arbitrary constant, we can find a continuity set B of μ such that

$$|\mu_j|(B^c) \leq \varepsilon \tag{2.6.9}$$

and

$$|\mu|(B^c) \leq \varepsilon. \tag{2.6.10}$$

If $\rho > 0$ is given, we can find some continuity sets $B_1, \ldots, B_m$ of μ such that

$$B = \bigcup_{k=1}^{m} B_k$$

and

$$C = \sup_{1 \leq k \leq m} \sup_{u,\, u' \in B_k} |u - u'| \leq \varepsilon/\rho.$$

If $\|t\| \leq \rho$,

$$\left| \int_{B_k} (e^{i(t,\, u_k)} - e^{i(t,\, u)}) \mu_j(du) \right| \leq C\rho\, |\mu_j|(B_k) < \varepsilon\, |\mu_j|(B_k)$$

for any $u_k \in B_k$, so that

$$\left| \int_B (\phi(t, u) - e^{i(t,\, u)}) \mu_j(du) \right| < \varepsilon \sup_j \|\mu_j\| \tag{2.6.11}$$

where

$$\phi(t, u) = \sum_{k=1}^{m} e^{i(t,\, u_k)} i_{B_k}(u).$$

In the same way, if $\|t\| \leq \rho$,

$$\left| \int_B (\phi(t, u) - e^{i(t,\, u)}) \mu(du) \right| \leq \varepsilon. \tag{2.6.12}$$

Moreover there exists some j_0 such that

$$\left| \int_B \phi(t, u)(\mu_j - \mu)(du) \right| \leq |\mu_j - \mu|(B) \leq \varepsilon \tag{2.6.13}$$

if $j > j_0$. From (2.6.11)–(2.6.13) and from

$$\begin{aligned}
&\left| \int_B e^{i(t,\, u)} \mu_j(du) - \int_B e^{i(t,\, u)} \mu(du) \right| \\
&\quad \leq \left| \int_B (\phi(t, u) - e^{i(t,\, u)}) \mu_j(du) \right| + \left| \int_B \phi(t, u)(\mu_j - \mu)(du) \right| \\
&\quad\quad + \left| \int_B (\phi(t, u) - e^{i(t,\, u)}) \mu(du) \right|,
\end{aligned}$$

we obtain

$$\left| \int_B e^{i(t,\, u)} (\mu_j - \mu)(du) \right| \leq \varepsilon \left(1 + \|\mu\| + \sup_j \|\mu_j\| \right) \tag{2.6.14}$$

if $j > j_0$ and $\|t\| \leq \rho$ and since

$$\left|\int e^{i(t,u)}\mu_j(du) - \int e^{i(t,u)}\mu(du)\right|$$
$$\leq \left|\int_B e^{i(t,u)}(\mu_j - \mu)(du)\right| + |\mu|(B^c) + |\mu_j|(B^c),$$

we have from (2.6.9), (2.6.10), and (2.6.14)

$$\left|\int e^{i(t,u)}\mu_j(du) - \int e^{i(t,u)}\mu(du)\right| \leq \varepsilon\left(3 + \|\mu\| + \sup_j\|\mu_j\|\right)$$

if $j > j_0$ and $\|t\| \leq \rho$ and this implies the theorem.

The converse which is false for arbitrary signed measures is true for measures. More precisely, we have

Theorem 2.6.8 Let $\{\mu_j\}$ be a sequence of finite measures. If $\{\hat{\mu}_j\}$ converges to a function f which is continuous at the origin, there exists a measure μ such that $\{\mu_j\}$ converges completely to μ and $\hat{\mu} = f$.

Proof We first prove

Lemma 2.6.5 If $\{\mu_j\}$ is a sequence of measures such that $\lim_{j\to\infty} \hat{\mu}_j(t) = f(t)$ is continuous at the origin, then $\{\mu_j\}$ is uniformly bounded and tight.

Proof Since

$$\hat{\mu}_j(t) = \|\mu_j\|$$

converges, $\{\|\mu_j\|\}$ is a bounded set of real numbers, so that $\{\mu_j\}$ is uniformly bounded. Since

$$\mu_j(\{x : \|x\| > m\sqrt{n}\}) \leq \sum_{k=1}^{n} \operatorname{pr}_{e_k} \mu_j(\{x : \|x\| > m\}),$$

it suffices to prove the tightness for univariate measures μ_j. Then, using Fubini's theorem

$$\begin{aligned}
\frac{1}{T}\int_{-T}^{T} (\hat{\mu}_j(0) - \hat{\mu}_j(t))\, dt &= \frac{1}{T}\int\left(\int_{-T}^{T}(1 - e^{itu})\, dt\right)\mu_j(du) \\
&= 2\int\left(1 - \frac{\sin(Tu)}{Tu}\right)\mu_j(du) \\
&\geq 2\int_{\{u:|u|\geq 2T\}}\left(1 - \frac{1}{T|u|}\right)\mu_j(du) \\
&\geq \mu_j\left(\left\{x : |x| \geq \frac{2}{T}\right\}\right).
\end{aligned}$$

If $\{\hat{\mu}_j\}$ converges to a function f which is continuous at the origin, then for any $\varepsilon > 0$ there exists some T such that

$$\frac{1}{T}\int_{-T}^{T} (f(0) - f(t))\, dt < \varepsilon$$

and, by the dominated convergence theorem, there exists some j_0 such that

$$\frac{1}{T}\int_{-T}^{T} (\hat{\mu}_j(0) - \hat{\mu}_j(t))\, dt < \varepsilon$$

for any $j > j_0$ and therefore

$$\mu_j(\{x : |x| > m\}) < \varepsilon$$

with $m = 2/T$ for $j > j_0$. This implies easily that $\{\mu_j\}$ is tight.

Proof of Theorem 2.6.8 From Helly's first theorem, we deduce the existence of a subsequence $\{\mu_{j_k}\}$ of $\{\mu_j\}$ which converges completely to some measure μ. From Theorem 2.6.7, we have

$$\lim_{k\to\infty} \hat{\mu}_{j_k} = \hat{\mu}$$

so that $\hat{\mu} = f$. From the uniqueness theorem, it follows that the limit μ is independent of the convergent subsequence $\{\mu_{j_k}\}$ and from Corollary 2.6.1, $\{\mu_j\}$ converges completely to μ.

Since the condition of uniform boundedness of measures is always satisfied by probabilities, Theorems 2.6.7 and 2.6.8 give

Theorem 2.6.9 (*Lévy's continuity theorem*) A sequence $\{p_j\}$ of n-variate probabilities converges completely to a probability p if and only if the sequence $\{\hat{p}_j\}$ converges to a function f which is continuous at the origin. Then $\hat{p} = f$.

2.7 BOCHNER'S THEOREM

We introduce the following definition. A function f defined on R^n with complex values is a *nonnegative-definite* function if it satisfies

$$\sum_{j=1}^{m} \sum_{k=1}^{m} \xi_j \bar{\xi}_k f(t_j - t_k) \geq 0 \tag{2.7.1}$$

for any $m \in N$, any complex numbers $\xi_1, \ldots, \xi_m$, and any $t_1, \ldots, t_m$ belonging to R^n.

The next theorem demonstrates the importance of this class of functions.

Theorem 2.7.1 (*Bochner's theorem*) f is a nonnegative definite function which is continuous at the origin if and only if there exists some measure μ such that $\hat{\mu} = f$.

Proof The necessity follows immediately from

$$\sum_{j=1}^{m} \sum_{k=1}^{m} \xi_j \overline{\xi_k} \hat{\mu}(t_j - t_k) = \sum_{j=1}^{m} \sum_{k=1}^{m} \xi_j \overline{\xi_k} \int e^{i(t_j - t_k, u)} \mu(du)$$
$$= \int \left| \sum_{j=1}^{m} \xi_j e^{i(t_j, u)} \right|^2 \mu(du) \geq 0.$$

Now, if f is a nonnegative-definite function, then

$$\phi_a(x) = \left(\prod_{j=1}^{n} a_j \right)^{-1} \int_{C_a \times C_a} f(u - v) e^{-i(u - v, x)} \lambda(du) \lambda(dv)$$

is nonnegative for any $x \in R^n$ and any $C_a = \{x \in R^n : 0 < x < a, a = (a_1, \ldots, a_n)\}$ since the integral in the right side is a limit of sums of the kind (2.7.1). A change of variables $(u, v) \to (t, u)$ with $t = u - v$ gives

$$\phi_a(x) = \int f_a(t) e^{-i(t, x)} \lambda(dt) \tag{2.7.2}$$

with

$$f_a(t) = \begin{cases} f(t) \prod_{j=1}^{n} (1 - (|t_j|/a_j)) & \text{if } -a < t < a, \\ 0 & \text{otherwise.} \end{cases}$$

If we consider the integral

$$I_b(u) = \frac{1}{(2\pi)^n} \int_{D_b} \prod_{j=1}^{n} \left(1 - \frac{|x_j|}{b}\right) \phi_a(x) e^{i(u, x)} \lambda(dx) \tag{2.7.3}$$

where $D_b = \{x \in R^n : -b < x_j < b, j = 1, \ldots, n\}$, then from (2.7.2) and Fubini's theorem, we obtain

$$I_b(u) = \frac{1}{(2\pi)^n} \int f_a(t) \left(\int_{D_b} \prod_{j=1}^{n} \left(1 - \frac{|x_j|}{b}\right) \phi_a(x) e^{i(u - t, x)} \lambda(dx) \right) \lambda(dt)$$
$$= \frac{1}{(2\pi)^n} \int f_a(t) \left(\prod_{j=1}^{n} \int_{-b}^{b} \left(1 - \frac{|x_j|}{b}\right) e^{i(u_j - t_j) x_j} \, dx_j \right) \lambda(dt)$$
$$= \frac{1}{\pi^n} \int f_a(t) \prod_{j=1}^{n} \frac{\sin^2(\frac{1}{2} b(t_j - u_j))}{b(\frac{1}{2}(t_j - u_j))^2} \lambda(dt)$$

and letting the change of variable $t = (2\tau/b) + u$,

$$I_b(u) = \frac{1}{\pi^n} \int f_a\left(u + \frac{2\tau}{b}\right) \prod_{j=1}^{n} \frac{\sin^2 \tau_j}{\tau_j^2} \lambda(d\tau). \tag{2.7.4}$$

Using the dominated convergence theorem, we find from (2.7.3)

$$\lim_{b \to \infty} I_b(u) = \frac{1}{(2\pi)^n} \int \phi_a(x) e^{i(u, x)} \lambda(dx)$$

and from (2.7.4) and

$$\int \frac{\sin^2 \tau_j}{\tau_j^2} d\tau_j = \pi,$$

$$\lim_{b \to \infty} I_b(u) = f_a(u),$$

so that f_a is the Fourier–Stieltjes transform of the measure having $(2\pi)^{-n}\phi_a$ as Radon–Nikodym derivative. Now letting $a \to \infty$, since f_a converges to f and since f is continuous at the origin, the theorem follows from Theorem 2.6.8.

We deduce easily from this theorem

Corollary 2.7.1 f is a nonnegative-definite function which is continuous at the origin and satisfies $f(0) = 1$ if and only if there exists a probability p having f for characteristic function.

There exist some other criteria for a function to be a characteristic function. In the case $n = 1$, they are given by Lukacs (1970, Section 4.2). Criteria for a function to be the Fourier–Stieltjes transform of a measure or a signed measure have been given by Riesz (1933), Bochner (1934), Schoenberg (1934), Haviland (1937), Cramér (1939), González Domínguez (1940), Yosida (1944), Kawata (1950), and Letta (1963).

2.8 A CHARACTERIZATION OF THE FOURIER–STIELTJES TRANSFORM

In this section, we show that some of the preceding properties characterize to some extent the Fourier–Stieltjes transform. We begin with

Lemma 2.8.1 A mapping f defined on $\mathcal{M}_n$ with values in $\mathcal{C}_C(R^n)$† satisfies the following conditions:

---

† $\mathcal{C}_F(E)$ denotes the set of all the continuous mappings defined on E with values in F.

(a) f is linear: $f(\alpha_1 \mu_1 + \alpha_2 \mu_2) = \alpha_1 f(\mu_1) + \alpha_2 f(\mu_2)$ for $\mu_j \in \mathscr{M}_n$ and $\alpha_j \in R$ $(j = 1, 2)$;

(b) f is bounded: there exists some constant K such that $\|\mu\| \leq 1$ implies $\sup_t |f(\mu)(t)| \leq K$;

(c) f is continuous: $\mu_j \xrightarrow{c} \mu$ implies $\lim_{j\to\infty} f(\mu_j)(t) = f(\mu)(t)$ for any $t \in R^n$;

(d) $f(\mu_1 * \mu_2) = f(\mu_1) f(\mu_2)$, if and only if $f = \hat{\mu} \circ \phi$ where $\phi \in \mathscr{C}_{R^n}(R^n)$.

Proof We first prove the necessity of the condition. If we denote by g_a the function $f(\delta_a)$ from (c), we deduce that g_a is continuous with respect to a and from (d) it follows that

$$f(\delta_{a+b}) = f(\delta_a * \delta_b) = f(\delta_a) f(\delta_b)$$

or

$$g_{a+b}(t) = g_a(t) g_b(t)$$

for any $a \in R^n$, $b \in R^n$, $t \in R^n$. From this, it follows [cf. Aczél (1966, p. 215)] that

$$g_a(t) = e^{(h(t), a)}$$

where $h \in \mathscr{C}_{C^n}(R^n)$. If $h = \psi + i\phi$ where $\phi \in \mathscr{C}_{R^n}(R^n)$, $\psi \in \mathscr{C}_{R^n}(R^n)$, then

$$|g_a(t)| = e^{(\psi(t), a)}$$

and if $\psi(t_0) \neq 0$ for some $t_0 \neq 0$, then

$$\lim_{l\to\infty} g_{l\psi(t_0)}(t_0) = +\infty$$

and f does not satisfy (b). Therefore

$$g_a(t) = e^{i(\phi(t), a)}$$

where $\phi \in \mathscr{C}_{R^n}(R^n)$.

Now if μ is a purely atomic signed measure having a finite number of atoms,

$$\mu = \sum_{j=1}^{p} \alpha_j \delta_{a_j} \qquad (\alpha_j \in R),$$

from (a) it follows that

$$f(\mu)(t) = \sum_{j=1}^{p} \alpha_j g_{a_j}(t) = \sum_{j=1}^{p} \alpha_j e^{i(\phi(t), a_j)} = \int e^{i(\phi(t), u)} \mu(du). \tag{2.8.1}$$

Now if $\mu \in \mathscr{M}_n$, there exists a sequence $\{\mu_j\}$ of purely atomic signed measures having a finite number of atoms which converges completely to μ. From (2.8.1) we have

$$f(\mu_j)(t) = \int e^{i(\phi(t),\, u)}\mu_j(du)$$

and from Theorem 2.6.7 and (c)

$$f(\mu)(t) = \int e^{i(\phi(t),\, u)}\mu(du)$$

for any $t \in R^n$, so that

$$f(\mu) = \hat{\mu} \circ \phi.$$

The converse follows immediately from the properties of Fourier–Stieltjes transforms.

Lemma 2.8.2 $f(\mu) = \hat{\mu} \circ \phi$ where $\phi \in \mathscr{C}_{R^n}(R^n)$ is injective if and only if $\mathrm{pr}_\theta \circ \phi(R^n)$ contains one of the sets

$$\{z \in R^n : z = \lambda\theta, \lambda > 0\} \qquad \text{or} \qquad \{z \in R^n : z = -\lambda\theta, \lambda > 0\}$$

for any $\theta \in R^n \backslash \{0\}$.

Proof We first prove the case $n = 1$ and suppose that $\phi(R)$ contains the set of all the positive numbers, the other case being identical. If

$$f(\mu_1) = f(\mu_2),$$

then

$$\hat{\mu}_1(t) = \hat{\mu}_2(t)$$

for any $t > 0$ and since a Fourier–Stieltjes transform is continuous and hermitian, we have $\hat{\mu}_1 = \hat{\mu}_2$ and, from the uniqueness theorem, $\mu_1 = \mu_2$.

Conversely, if ϕ does not satisfy the condition of the lemma, there exists some $a \in R$ such that $\phi(R)$ does not contain either $[-a, +a]$ or $\{x : |x| > a\}$. From Pólya's theorem (cf. Theorem 3.1.1), we deduce the existence of distinct probabilities p_1 and p_2 such that

$$\hat{p}_1(t) = \hat{p}_2(t)$$

for any $t \in \phi(R)$ and f is not injective.

We now prove the general case. If ϕ satisfies the condition for any $\theta \in R^n$ and if $f(\mu_1) = f(\mu_2)$, it follows from the case $n = 1$ that

$$\mathrm{pr}_\theta\, \mu_1 = \mathrm{pr}_\theta\, \mu_2$$

for any $\theta \in R^n$ and, from Theorem 2.4.2, $\mu_1 = \mu_2$. If ϕ does not satisfy the condition for some $\theta \in R^n$, from the case $n = 1$, we deduce the existence of distinct probabilities p_1 and p_2 concentrated on the one-dimensional subspace generated by θ and such that $f(p_1) = f(p_2)$. Therefore f is not injective.

Lemma 2.8.3 Let $f(\mu) = \hat{\mu} \circ \phi$ where $\phi \in \mathscr{C}_{R^n}(R^n)$. Then $\phi(t) = Ct$ for some real constant C if and only if f satisfies the two conditions:

(a) $f(\delta_a)$ is a nonnegative-definite function for any $a \in R^n$;
(b) $f(\mu)(\mathrm{pr}_\theta\, t) = f(\mathrm{pr}_\theta\, \mu)(t)$ for any $t \in R^n$.

Proof If

$$f(\delta_a)(t) = e^{i(\phi(t),\, a)}$$

is a nonnegative-definite function for any $a \in R^n$, then by definition

$$\sum_{j=1}^{m} \sum_{k=1}^{m} \xi_j \overline{\xi_k} e^{i(\phi(t_j - t_k),\, a)} \geq 0$$

for any $m \in N$, $\xi_j \in C$, $t_j \in R^n$ $(j = 1, \ldots, m)$. Letting $m = 1$, $\xi_1 = 1$, we obtain

$$e^{i(\phi(0),\, a)} \geq 0$$

for any $a \in R^n$, so that $\phi(0) = 0$ and

$$f(\delta_a)(0) = 1$$

for any $a \in R^n$. From Corollary 2.7.1, we deduce the existence of a probability p_a such that

$$f(\delta_a)(t) = e^{i(\phi(t),\, a)} = \hat{p}_a(t).$$

Since

$$f(\delta_{-a})(t) = e^{-i(\phi(t),\, a)} = (\hat{p}_a(t))^{-1} = \hat{p}_{-a}(t),$$

$\hat{p}_{-a}$ is the reciprocal of $\hat{p}_a$. But, from Theorem 2.5.7, it follows easily that the only characteristic functions whose reciprocals are also characteristic functions are the characteristic functions of degenerate probabilities, so that there exists a function $m \in \mathscr{C}_{R^n}(R^n)$ such that

$$e^{i(\phi(t),\, a)} = e^{i(t,\, m(a))}$$

for any $t \in R^n$, $a \in R^n$. This is equivalent to

$$(\phi(t), a) = (t, m(a)) + 2k\pi$$

where k is a continuous function with integer values. Letting $a = t = 0$, we obtain $k = 0$.

Now, if $t = (t_1, \ldots, t_n)$, $a = (a_1, \ldots, a_n)$, we obtain from (b)

$$f(\delta_{a_j e_j})(t) = f(\delta_a)(t_j e_j),$$

so that

$$(\phi(t), a_j e_j) = (t_j e_j, m(a)),$$

that is,

$$a_j \operatorname{pr}_{e_j} \phi(t) = t_j \operatorname{pr}_{e_j} m(a).$$

Therefore $\operatorname{pr}_{e_j} \phi(t)$ depends only on t_j and

$$(\operatorname{pr}_{e_j} \phi(t))/t_j = (\operatorname{pr}_{e_j} m(a))/a_j$$

is a real constant C_j and we obtain

$$\phi(t_1, \ldots, t_n) = (C_1 t_1, \ldots, C_n t_n).$$

In the same way, we have

$$f(\delta_{e_j})(\operatorname{pr}_{e_j+e_k} t) = f(\operatorname{pr}_{e_j+e_k} \delta_{e_j})(t).$$

Since

$$\operatorname{pr}_{e_j+e_k} \delta_{e_j} = \delta_{\frac{1}{2}\sqrt{2}(e_j+e_k)},$$

we have

$$f(\operatorname{pr}_{e_j+e_k} \delta_{e_j})(t) = \exp(\tfrac{1}{2}i\sqrt{2}\,(C_j t_j + C_k t_k))$$

while

$$f(\delta_{e_j})(\operatorname{pr}_{e_j+e_k} t) = \exp(\tfrac{1}{2}iC_j\sqrt{2}\,(t_j + t_k)),$$

so that

$$C_j = C_k$$

if $j \neq k$ and the necessity of the condition is proved. Its sufficiency follows immediately from the properties of the Fourier–Stieltjes transforms.

From these three lemmas, we obtain easily the following.

Theorem 2.8.1 A mapping f defined on $\mathcal{M}_n$ with values on $\mathcal{C}_C(R^n)$ is of the kind

$$f(\mu)(t) = \int e^{iC(t, x)}\mu(dx)$$

for some real constant C different from zero if and only if f satisfies the following conditions:

(a) f is linear;
(b) f is injective;
(c) f is bounded;
(d) f is continuous;

(e) $f(\mu_1 * \mu_2) = f(\mu_1) f(\mu_2)$;
(f) $f(\mu)(\mathrm{pr}_\theta\, t) = f(\mathrm{pr}_\theta\, \mu)(t)$ for any $t \in R^n$;
(g) $f(\delta_a)$ is a nonnegative-definite function for any $a \in R^n$.

2.9 EXPONENTIAL OF SIGNED MEASURES

If $\mu \in \mathscr{M}_n$, we call the *exponential* of μ the signed measure $\exp \mu$ defined by

$$\exp \mu = \sum_{j=0}^{\infty} (1/j!)\mu^{*j} \tag{2.9.1}$$

where μ^{*j} is defined inductively for $j \in N$ by

$$\mu^{*0} = \delta_0, \qquad \mu^{*j} = \mu^{*(j-1)} * \mu \qquad \text{if} \quad j > 0.$$

The series in (2.9.1) must be defined as the complete limit of

$$\sum_{j=0}^{m} (1/j!)\mu^{*j}$$

when $m \to \infty$ (the existence of this limit follows immediately from Corollary 2.5.4). We have also

$$\|\exp \mu\| \leq e^{\|\mu\|}.$$

If μ is concentrated on $A \in F_\sigma^n$, $\exp \mu$ is concentrated on $(N)A$.

From Theorems 2.5.4 and 2.6.7, we have immediately

Theorem 2.9.1 If $\mu \in \mathscr{M}_n$, $\exp \hat{\mu} = \widehat{\exp \mu}$.

Corollary 2.9.1 If $\mu_j \in \mathscr{M}_n$ $(j = 1, 2)$, then $\exp \mu_1 = \exp \mu_2$ if and only if $\mu_1 = \mu_2$.

Corollary 2.9.2 If $\mu_j \in \mathscr{M}_n$ $(j = 1, 2)$, then

$$\exp(\mu_1 + \mu_2) = \exp \mu_1 * \exp \mu_2 .$$

Corollary 2.9.3 If $\mu \in \mathscr{M}_n$, $\exp \mu$ is concentrated on a subspace E of R^n if and only if μ is concentrated on E.

Corollary 2.9.4 If $\mu_j \in \mathscr{M}_n$ $(j = 1, 2, \ldots)$ and if $\{\mu_j\}$ converges completely to μ, then

$$\lim_{j \to \infty} \exp \hat{\mu}_j = \exp \hat{\mu}.$$

Corollary 2.9.5 If μ_j is a finite measure $(j = 1, 2, \ldots)$, then $\{\mu_j\}$ converges completely to μ if and only if $\{\exp \mu_j\}$ converges completely to $\exp \mu$.

These results follow immediately from Theorem 2.9.1 and Theorem 2.2.2, Theorem 2.5.4, Corollary 2.4.1, Theorems 2.6.7 and 2.6.8, respectively.

Let E_1 and E_2 be two subspaces of R^n such that

$$E_1 \oplus E_2 = R^n$$

and μ_j be a finite signed measure concentrated on E_j ($j = 1, 2$). We define the *direct product* of the μ_j as the convolution $\mu = \mu_1 * \mu_2$ of these signed measures and *direct sum* of the μ_j as the sum $\nu = \mu_1 + \mu_2$ of these signed measures. We write

$$\mu = \mu_1 \otimes \mu_2 , \qquad \nu = \mu_1 \oplus \mu_2 .$$

Theorem 2.9.2 Let $\mu \in \mathscr{M}_n$. Then

$$\exp \mu = \nu_1 \otimes \nu_2 \tag{2.9.2}$$

if and only if

$$\nu_j = \exp \mu_j \qquad (j = 1, 2)$$

with

$$\mu = \mu_1 \oplus \mu_2 .$$

Proof The sufficiency follows from Corollaries 2.9.2 and 2.9.3. We prove now the necessity. From (2.9.2), we have

$$\exp \hat{\mu}(t) = \hat{\nu}_1(\mathrm{pr}_{E_1} t)\hat{\nu}_2(\mathrm{pr}_{E_2} t).$$

Letting $\mathrm{pr}_{E_2} t = 0$, we obtain

$$\hat{\nu}_1(\mathrm{pr}_{E_1} t) = \exp(\hat{\mu}(\mathrm{pr}_{E_1} t)),$$

so that

$$\nu_1 = \exp \mu_1$$

with $\mu_1 = \mathrm{pr}_{E_1} \mu$. In the same way, we prove that

$$\nu_2 = \exp \mu_2$$

with $\mu_2 = \mathrm{pr}_{E_2} \mu$. From

$$\exp \mu = \exp \mu_1 * \exp \mu_2 ,$$

we have from Corollary 2.9.2

$$\exp \mu = \exp(\mu_1 + \mu_2)$$

and since from Corollary 2.9.3, μ_j is concentrated on E_j ($j = 1, 2$), we obtain from Corollary 2.9.1

$$\mu = \mu_1 \oplus \mu_2 .$$

NOTES

The first use of integral transforms in probability theory is due to Laplace (1795) who introduced the "generating function" $\int x^s p(dx)$ for purely atomic probabilities having for atoms the nonnegative integers. Then, Cauchy (1853a,b,c) used the characteristic function of some particular absolutely continuous probabilities. Later, Poincaré (1912) introduced the transform $\int e^{tx} p(dx)$, but this transform is not always defined. Finally, Lévy gave in 1923 the definition and the main properties of characteristic functions of univariate probabilities. His work during this period can be found in Lévy (1925). Later, Bochner (1933) proved the so-called *Bochner's theorem* (even in the case $n > 1$).

The study of multivariate characteristic functions is historically more intricate, but important contributions are due to Wintner (1934) and Haviland (1934a,b, 1935a,b). The projection theorem has been proved by Cramér and Wold (1936). Good expositions of the classical theory are given by Cramér (1937), Lévy (1937b), Wintner (1938, 1947), and Gnedenko (1944).

The characterization of Fourier–Stieltjes transformation by its properties (Theorem 2.8.1) is due in the case $n = 1$ to Lukacs (1964) [see also Lukacs (1952) for a less complete result].

Chapter 3

Analytic Characteristic Functions

3.1 EXAMPLES OF CHARACTERISTIC FUNCTIONS

In this section, we give three examples of multivariate probabilities which are useful in the following.

Example 1 *Degenerate probabilities* If δ_α is the degenerate probability defined in Section 1.1, then

$$\hat{\delta}_\alpha(t) = e^{i(\alpha, t)} \qquad (t \in R^n).$$

Example 2 *Normal probabilities* If $\alpha \in R^n$ and Q is a nonnegative quadratic form, then the function f defined by

$$f(t) = \exp(i(\alpha, t) - Q(t)) \qquad (t \in R^n) \tag{3.1.1}$$

is the characteristic function of a normal probability $\mathfrak{n}(\alpha, Q)$. If Q is positive, then f is integrable and, from Corollary 2.3.1, $\mathfrak{n}(\alpha, Q)$ is absolutely continuous. If there exists some $t_0 \neq 0$ such that $Q(t_0) = 0$, then, from the properties of quadratic forms, it can be shown that the support of $\mathfrak{n}(\alpha, Q)$ is of the kind $E + \alpha$ where E is a proper subspace of R^n. From the projection theorem, we deduce that all the projections of $\mathfrak{n}$ are normal probabilities. We will prove in Section 3.3 that the converse is true.

Example 3 *Poisson probabilities* If f is the function

$$f(t) = \exp\left(\sum_{\varepsilon} l_{\varepsilon}(e^{i(\varepsilon, t)} - 1)\right) \qquad (t \in R^n) \tag{3.1.2}$$

where $\varepsilon = (\varepsilon_1, \ldots, \varepsilon_n)$, $\varepsilon_j = 0$ or 1, $\sum_{\varepsilon}$ means the summation on the $2^n - 1$ values of $\varepsilon \neq 0$ and the l_{ε} are $2^n - 1$ nonnegative constants, then f is the characteristic function of a probability p concentrated on N^n and all the margins of p are univariate Poisson probabilities. Consequently, Teicher (1954) has called this probability an *n-variate Poisson probability*. Lévy (1968) has criticized this definition, arguing that this property depends on the choice of coordinate axis and he proposed for definition of an n-variate Poisson probability q the following one:

$$\hat{q}(t) = \exp(l(e^{i(\varepsilon, t)} - 1)) \qquad (t \in R^n) \tag{3.1.3}$$

where $\varepsilon = (\varepsilon_1, \ldots, \varepsilon_n)$ with $\varepsilon_j = 0$ or 1 $(j = 1, \ldots, n)$ and l is a positive constant. q has the property that all its univariate projections are Poisson-type probabilities. Nevertheless, since (3.1.3) can be considered as a particular case of (3.1.2) and since the function (3.1.2) is more useful for decomposition theorems, we conserve Teicher's definition and define an n-variate Poisson probability as the probability having (3.1.2) as characteristic function and more generally Poisson-type probability as the probability $\mathfrak{p}(\alpha, m, l)$ $[\alpha \in R^n,\ \ m = (m_1, \ldots, m_n) \in R^n,\ \ l = (l_{1, 0, \ldots, 0},\ \ l_{0, 1, 0, \ldots, 0}, \ldots, l_{1, \ldots, 1}) \in (R^+)^{2^n - 1}]$ having for characteristic function

$$\hat{\mathfrak{p}}(\alpha, m, l)(t) = \exp\left(i(\alpha, t) + \sum_{\varepsilon} l_{\varepsilon}\left(\exp\left(i \sum_{j=1}^{n} \varepsilon_j m_j t_j\right) - 1\right)\right). \tag{3.1.4}$$

Evidently the probability $\mathfrak{p}(\alpha, m, l)$ defined by (3.1.4) is a lattice probability with span m. When $\alpha = 0$, we denote by $\mathfrak{p}(m, l)$ the probability $\mathfrak{p}(0, m, l)$.

For other examples of characteristic functions in the case $n = 1$, see Lukacs (1970, p. 18) and in the general case, see Lukacs and Laha (1964, Chapter 2).

We have also the two following results [for the proof, see Lukacs (1970, Section 4.3)].

Theorem 3.1.1 (*Pólya's theorem*) If a continuous function f defined on the real numbers with real values satisfies the following conditions:

(a) $f(0) = 1$,
(b) f is even,
(c) f is convex for $t > 0$,
(d) $\lim_{t \to \infty} f(t) = 0$,

then f is the characteristic function of an absolutely continuous probability.

Theorem 3.1.2 (*Dugué–Girault theorem*) If a function f defined on the real numbers with real values satisfies the following conditions:

(a) $f(0) = 1$,
(b) f is even,
(c) f is convex and continuous on $[0, r]$,
(d) f is periodic with period $2r$,
(e) $f(r) = 0, f(t) > 0$ on $[0, r]$,

then f is the characteristic function of a lattice probability.

We do not know the extension of these two results to the case $n > 1$.

3.2 DERIVATIVES OF CHARACTERISTIC FUNCTIONS

If $p \in \mathscr{P}_n$ and $k \in N^n$, the number α_k defined by

$$\alpha_k = \int u^k p(du)$$

is the moment of order k of p if the integral exists and the number β_k defined by

$$\beta_k = \int |u|^k p(du)$$

is the absolute moment of order k of p if the integral exists. If one of the integrals does not exist, the corresponding moment of p does not exist. From the properties of Lebesgue–Stieltjes integrals, we have immediately

Theorem 3.2.1 If $p \in \mathscr{P}_n$, the moment α_k of order k of p exists if and only if the absolute moment β_k of order k of p exists. Moreover

$$|\alpha_k| \leq \beta_k \qquad \text{and} \qquad \alpha_{2k} = \beta_{2k}.$$

In the case $n = 1$, it is easy to see that the existence of a moment of order k implies the existence of moments of order k' for any $k' \leq k$. This result is not true in the case $n > 1$. For example, if p is concentrated on the first coordinate axis, then $\alpha_k = 0$ for any $k \neq me_1$ $(m \in N)$, but the existence of α_{me_1} implies some additional conditions on p. Nevertheless, we have

Theorem 3.2.2 Let $m \in N$ and $p \in \mathscr{P}_n$. If the moment of order me_j of p exists $(j = 1, \ldots, n)$, then the moment of order k of p exists for any $k = (k_1, \ldots, k_n) \in N^n$ such that

$$\sum_{j=1}^{n} k_j = m.$$

Proof From Hölder's inequality [see Hardy *et al.* (1952, p. 26)]

$$\left(\sum_{j=1}^{n} |u_j|\right)^m \leq n^{m-1} \sum_{j=1}^{n} |u_j|^m,$$

we have

$$|u|^k \leq n^{m-1} \sum_{j=1}^{n} |u|^{me_j},$$

and this implies the theorem.

Theorem 3.2.3 Let $k \in N^n$ and $\varepsilon = (\varepsilon_1, \ldots, \varepsilon_n)$ with $\varepsilon_j = 0$ or 1 $(k \geq \varepsilon)$. If the moments $\alpha_{2(k-\varepsilon)}$ and α_{2k} of a probability p exist, then the moment $\alpha_{2k-\varepsilon}$ exists and

$$\alpha_{2k-\varepsilon} \leq \tfrac{1}{2}(\alpha_{2k} + \alpha_{2(k-\varepsilon)}).$$

Proof This follows immediately from the well-known inequality

$$|l| \leq \tfrac{1}{2}(1 + l^2)$$

valid for any real l.

We can derive many results on the links between the moments of a probability p and the derivatives of $\hat{p}$ [for a study in the case $n = 1$, see Lukacs (1970, Section 2.3)]. Here we give only the simplest of these results which are also the most useful.

Theorem 3.2.4 Let $p \in \mathscr{P}_n$ and $k \in N^n$. If the moments of order k' of p exist for any $k' \leq k$, then $D^k\hat{p}$ exists and

$$D^k\hat{p}(t) = \int (iu)^k e^{i(t,u)} p(du).$$

Proof If the moment of order k' exists, then the integral

$$\int (iu)^{k'} e^{i(t,u)} p(du)$$

is absolutely and uniformly convergent for any $t \in R^n$, so that $\hat{p}$ can be derived k times under the integration sign.

From this theorem, we deduce immediately the following corollary.

Corollary 3.2.1 With the conditions of Theorem 3.2.4,

$$D^k\hat{p}(0) = i^{k_1 + \cdots + k_n}\alpha_k, \qquad |D^k\hat{p}(t)| \leq \beta_k$$

for any $t \in R^n$ where α_k and β_k are, respectively, the moment and the absolute moment of order k of p.

Theorem 3.2.5 Let $p \in \mathscr{P}_n$ and $k \in N^n$. If $D^{2k}\hat{p}(0)$ exists, then the moment α_{2k} of order $2k$ of p exists and if $\alpha_{2k} \neq 0$, $\alpha_{2k}^{-1} D^{2k}\hat{p}$ is a characteristic function.

Proof By induction, it suffices to prove this result for $k = e_j$. Then

$$\begin{aligned} D^{2e_j}\hat{p}(0) &= \lim_{l \to 0} \frac{\hat{p}(2le_j) + \hat{p}(-2le_j) - 2}{4l^2} \\ &= -\lim_{l \to 0} \int \left(\frac{\sin(l(e_j, u))}{l} \right)^2 p(du) \\ &= -\int (e_j, u)^2 p(du) = -\alpha_{2e_j} \end{aligned}$$

and, if $\alpha_{2e_j} \neq 0$, $\alpha_{2e_j}^{-1} D^{2e_j}\hat{p}$ is the characteristic function of the probability q_j defined by

$$q_j(B) = \alpha_{2e_j}^{-1} \int_B u^{2e_j} p(du)$$

for any $B \in \mathscr{B}_n$.

This result is false if the order of the derivative is not even [for an example, see Lukacs (1970, p. 22)].

From Theorems 3.2.5 and 3.2.3, we deduce immediately

Corollary 3.2.2 With the conditions of Theorem 3.2.5, the moments of order k' of p exist for any $k' \leq 2k$.

3.3 ANALYTIC CHARACTERISTIC FUNCTIONS

If $p \in \mathscr{P}_n$, $\hat{p}$ is analytic if there exists a function f defined on C^n with complex values which is regular in some neighborhood of the origin and a positive constant δ such that

$$\hat{p}(t) = f(t) \qquad (t \in R^n, \quad \|t\| < \delta). \tag{3.3.1}$$

If f is an entire function, we say that $\hat{p}$ is an *entire characteristic function.* For example, if p is a normal or a Poisson probability, $\hat{p}$ is an entire characteristic function.

Let now $\hat{p}$ be an analytic characteristic function and f be a function which is regular in some neighborhood of the origin and satisfies (3.3.1). If α_k and β_k are, respectively, the moment and the absolute moment of order k of p, we have then from Corollaries 3.2.1 and 3.2.2

$$f(z) = \sum_k (\alpha_k/k!)(iz)^k. \tag{3.3.2}$$

By assumption, the series in the right side of (3.3.2) converges for $z \in D_{0,r}$ ($r \in R^n, r > 0$).† If we consider the series

$$\sum_k (\alpha_{2k}/(2k)!)(iz)^{2k} \tag{3.3.3}$$

and

$$\sum_k (\alpha_k/k!)(iz)^k - \sum_k (\alpha_{2k}/(2k)!)(iz)^{2k}, \tag{3.3.4}$$

they are extracted from the series (3.3.2) and consequently they converge, respectively, in $D_{0,r'}$ and $D_{0,r''}$ with $r' \geq r$ and $r'' \geq r$. From Theorem 3.2.3, we deduce easily that the convergence of (3.3.3) implies the convergence of (3.3.4) so that $r'' \geq r'$ and $r' = r$. In the same way, we can prove the convergence of the series

$$\sum_k (\beta_k/k!)z^k \tag{3.3.5}$$

in $D_{0,r}$.

Let now $\xi \in R^n$. We can define the Taylor series

$$\sum_k (1/k!)D^k\hat{p}(\xi)(z - \xi)^k$$

and, from Corollary 3.2.1, we deduce that this series converges in $D_{\xi,r}$. Moreover, from Theorem 3.2.4, we have

$$f(z) = \sum_k (1/k!)\left(\int (iu)^k e^{i(\xi,u)}p(du)\right)(z - \xi)^k.$$

From Corollary 3.2.1 and the convergence of (3.3.5), we deduce that we can exchange the order of summation and integration, so that

$$f(z) = \int \left(\sum_k (1/k!)(iu)^k(z - \xi)^k\right)e^{i(\xi,u)}p(du) = \int e^{i(z,u)}p(du).$$

More generally, if f is regular in $D_{iy_0,r}$ ($y_0 \in R^n$, $r \in R^n$, $r > 0$) and admits in some neighborhood of iy_0 the representation

$$f(z) = \int e^{i(z,u)}p(du), \tag{3.3.6}$$

† If $\xi = (\xi_1, \ldots, \xi_n) \in C^n$ and $r = (r_1, \ldots, r_n) \in R^n$, $D_{\xi,r}$ denote the polydisk with center ξ and polyradius r such that

$$D_{\xi,r} = \{x = (x_1, \ldots, x_n) : |x_j - \xi_j| < r_j, j = 1, \ldots, n\}.$$

we show in the same way that f is regular in $D_{\xi + iy_0, r}$ for any $\xi \in R^n$ and that in $\bigcup_\xi D_{\xi + iy_0, r}$ we have (3.3.6).

Now, if the integral

$$\int e^{-(y, u)} p(du) \tag{3.3.7}$$

converges for any y belonging to some neighborhood of y_0, it follows easily that (3.3.6) is regular at the point iy_0. Therefore f is regular in $R^n + i\Gamma$ where Γ is the interior of the domain of convergence of the integral (3.3.7). Γ is evidently a convex set of R^n.

Let now u be a point of the boundary of Γ. If iu is a regular point of f, there exists some $D_{iy_0, r}$ ($y_0 \in R^n$, $r \in R^n$, $r > 0$) where f is regular and such that $iu \in D_{iy_0, r}$. But (3.3.7) converges for $\|y - y_0\| < r$ and this contradicts the definition of Γ as the greatest open set of R^n where the integral (3.3.7) converges.

If we use the definition

a set $R^n + i\Gamma = \{z \in C^n : \text{Im } z \in \Gamma\}$ is a tube of C^n; if Γ is convex, $R^n + i\Gamma$ is a convex tube,

we have proved the following result.

Theorem 3.3.1 If a characteristic function $\hat{p}$ is analytic, it is regular in a convex tube $R^n + i\Gamma$ and admits in this tube the representation†

$$\hat{p}(z) = \int e^{i(z, u)} p(du). \tag{3.3.8}$$

Γ is the interior of the set $\{y \in R^n : \int e^{-(y, u)} p(du) < +\infty\}$. If y_0 is a point of the boundary of Γ, then iy_0 is a singular point of $\hat{p}$.

Corollary 3.3.1 (*Ridge property*) If $\hat{p}$ is an analytic characteristic function in a convex tube $R^n + i\Gamma$, then

$$|\hat{p}(x + iy)| \leq \hat{p}(iy)$$

for any $x \in R^n$, $y \in \Gamma$. More generally,

$$|D^{2k}\hat{p}(x + iy)| \leq |D^{2k}\hat{p}(iy)|$$

for any $k \in N^n$, $x \in R^n$, $y \in \Gamma$.

Proof From representation (3.3.8), we have

$$D^k\hat{p}(z) = \int (iu)^k e^{i(z, u)} p(du) \qquad (\text{Im } z \in \Gamma)$$

for any $k \in N^n$ and this implies the corollary.

---

† Here and in the following, we use the same notation for an analytic characteristic function and its regular extension.

Let now D be a domain of R^n (or C^n). If $\hat{p}$ has no zeros in D, we can define $\log \hat{p}$ by continuity from the value $\log \hat{p}(y_0)$ for some point $y_0 \in D$. [For a simple proof of this assertion, see Chung (1968, Theorem 7.6.2)]. In the following, we suppose always that $0 \in D$ and we denote by $\log \hat{p}$ the branch of the logarithm defined by continuity from $\log \hat{p}(0) = 0$. We have then from this definition and from Corollary 3.3.1

Corollary 3.3.2 If $\hat{p}$ is an analytic characteristic function without zeros in a convex tube $R^n + i\Gamma$, then

$$\text{Re}[\log \hat{p}(iy) - \log \hat{p}(x + iy)] \geq 0$$

for any $x \in R^n$, $y \in \Gamma$.

We give now a converse of Theorem 3.3.1.

Theorem 3.3.2 If Γ is a subset of R^n which is open, convex, and which contains the origin, a characteristic function $\hat{p}$ is analytic in $R^n + i\Gamma$ if and only if the integral $\int e^{-(y,u)}p(du)$ converges for any $y \in \Gamma$.

Proof The necessity follows immediately from Theorem 3.3.1. To prove the sufficiency, let $y_0 \in \Gamma$ and ρ be a constant lesser than the distance between y_0 and the boundary of Γ. Then there exists some constant M such that

$$\int e^{-(y,u)}p(du) \leq M$$

for any y such that $\|y - y_0\| \leq \rho$. Then the integral $\int e^{i(z,x)}p(dx)$ converges absolutely and uniformly and therefore is a regular function in $\{z \in C^n : \|\text{Im } z - y_0\| \leq \rho\}$. The theorem follows immediately from this assertion.

From this theorem, we deduce easily the following corollaries.

Corollary 3.3.3 Let Γ be the set defined in Theorem 3.3.2. Then a characteristic function $\hat{p}$ is analytic in $R^n + i\Gamma$ if and only if its projection $\text{pr}_\theta \hat{p}$ is analytic in the tube $\{z \in C^n : -j(-\theta) < \text{pr}_\theta \text{ Im } z < j(\theta)\}$ for any $\theta \in R^n$ where j is the gauge of Γ,†

$$j(\theta) = \sup\{\lambda \in R : \lambda\theta \in \Gamma\}.$$

Corollary 3.3.4 Let $\{f_1, \ldots, f_n\}$ be a basis of R^n and α_j and β_j $(j = 1, \ldots, n)$ be some positive constants. If $\text{pr}_{f_j} \hat{p}$ is regular in $\{z \in C^n : -\alpha_j < \text{pr}_{f_j} \text{Im } z < \beta_j\}$ $(j = 1, \ldots, n)$, then $\hat{p}$ is regular in $R^n + i\Gamma$ where Γ is the smallest convex set containing the n segments $\{x = (x_1, \ldots, x_n) : x = \lambda f_j, -\alpha_j < \lambda < \beta_j\}$.

† In general, the analysts call *gauge* the reciprocal of the function given here, but our definition is more convenient for our purposes.

Corollary 3.3.5 Let Γ be the set defined in Theorem 3.3.2 and $y \in \Gamma$. If $\hat{p}$ is a characteristic function regular in $R^n + i\Gamma$, then the function f_y defined by

$$f_y(z) = \hat{p}(z + iy)/\hat{p}(iy)$$

is a characteristic function which is regular in $R^n + i(\Gamma - y)$.

Proof The measure p_y defined by

$$p_y(B) = (\hat{p}(iy))^{-1} \int_B e^{-(y,\, u)} p(du) \tag{3.3.9}$$

for any $B \in \mathscr{B}_n$ is a probability and $\hat{p}_y = f_y$. The analyticity of f_y follows immediately from Theorem 3.3.2.

Theorem 3.3.3 Let Γ be the set defined in Theorem 3.3.2. Then a characteristic function $\hat{p}$ is analytic in $R^n + i\Gamma$ and is not analytic in any greater convex tube if and only if

$$j(\theta) = -\limsup_{u \to \infty} (1/u) \log \hat{p}(\{x : (x, \theta) \le -u\})$$

for any $\theta \in R^n$ where j is the gauge of Γ. [If for some $u > 0$, $p(\{x : (x, \theta) \le -u\}) = 0$, then the limit in the second member is by convention $-\infty$].

Proof In the case $n = 1$, this result is a consequence of Corollary 5 to Theorem 11.1.1 of Lukacs (1970) [see also Ramachandran (1967, Theorem 2.1.3)]. The general case follows from this one and from Corollary 3.3.3.

We can deduce with the same method some results on entire multivariate characteristic functions from the corresponding theorems on entire univariate characteristic functions [see Lukacs (1970, Section 7.2) or Ramachandran (1967, Chapter 3)]. We give only one example which follows from Theorem 7.2.5 of Lukacs (1970):

Theorem 3.3.4 $\hat{p}$ is an entire characteristic function of order $1 + \alpha^{-1}$ $(\alpha > 0)$ if and only if the two following conditions are satisfied:

(a) $p(\{x : \|x\| > u\}) > 0$ for any $u > 0$;

(b) $$\liminf_{u \to \infty} \frac{\log \log[p(\{x : \|x\| > u\})]^{-1}}{\log u} = 1 + \alpha.$$

We prove now that Theorem 3.3.1 is the best possible by showing that for any convex tube $R^n + i\Gamma$, Γ being open and containing the origin, there exists a characteristic function having this tube for domain of analyticity.†

† A domain $\Gamma \subset C^n$ is the domain of analyticity of a function f if f is regular in Γ and if each point of $\partial\Gamma$ is a singular point of f.

We first study the case $n = 1$ and prove even more in this case with

Theorem 3.3.5 If the domain $D \subset C$ satisfies the following conditions:

(a) D contains a strip $\{z \in C : -\alpha < \operatorname{Im} z < \beta\}$ $(0 < \alpha \leq +\infty,$ $0 < \beta \leq +\infty)$,

(b) D is symmetric with respect to the imaginary axis (if $(a, b) \in R^2$ is such that $a + ib \in D$, then $-a + ib \in D$),

(c) if $\beta < +\infty$ (resp. $\alpha < +\infty$), the point $i\beta$ (resp. $-i\alpha$) belongs to the boundary ∂D of D,

there exists an univariate characteristic function having D for domain of analyticity.

For the proof, we will use

Lemma 3.3.1 If α, β, and γ are real constants satisfying $0 < \alpha \leq \beta$, $\gamma \geq 0$, the function f defined by

$$f(t) = \left[\left(1 - \frac{it}{\alpha}\right)\left(1 - \frac{it}{\beta + i\gamma}\right)\left(1 - \frac{it}{\beta - i\gamma}\right)\right]^{-1}$$

is an univariate characteristic function.

Proof If $\gamma > 0$, we have

$$f(t) = a\left(1 - \frac{it}{\alpha}\right)^{-1} + b\left(1 - \frac{it}{\beta + i\gamma}\right)^{-1} + \bar{b}\left(1 - \frac{it}{\beta - i\gamma}\right)^{-1} \tag{3.3.10}$$

with

$$a = \left[\left(1 - \frac{\alpha}{\beta + i\gamma}\right)\left(1 - \frac{\alpha}{\beta - i\gamma}\right)\right]^{-1} = \frac{\beta^2 + \gamma^2}{(\beta - \alpha)^2 + \gamma^2} > 0,$$

$$b = \left[\left(1 - \frac{\beta + i\gamma}{\alpha}\right)\left(1 - \frac{\beta + i\gamma}{\beta - i\gamma}\right)\right]^{-1}$$

Now, let ϕ be the function defined by

$$\phi(x) = (1/2\pi)\int_{-\infty}^{\infty} f(t)e^{-itx}\, dt. \tag{3.3.11}$$

From (3.3.10), we obtain

$$\phi(x) = \begin{cases} a\alpha e^{-\alpha x} + b(\beta + i\gamma)e^{-(\beta + i\gamma)x} + \bar{b}(\beta - i\gamma)e^{-(\beta - i\gamma)x} & \text{if } x > 0, \\ 0 & \text{if } x < 0. \end{cases}$$

Since

$$b(\beta + i\gamma) = -\frac{a\alpha}{2}\left(1 + \frac{i(\beta - \alpha)}{\gamma}\right),$$

we have

$$\phi(x) = \alpha\alpha e^{-\alpha x}\left[1 - e^{-(\beta-\alpha)x}\left(\cos(\gamma x) + \frac{\beta - \alpha}{\gamma}\sin(\gamma x)\right)\right] \geq 0$$

if $x > 0$ and f is the characteristic function of the measure having ϕ for its Radon–Nikodym derivative. If $\gamma = 0$, the result follows immediately from the preceding one and from continuity theorem.

Proof of Theorem 3.3.5 If $\alpha = \beta = +\infty$, then D is the whole complex plane and we have many examples of entire characteristic functions. We suppose now that $\alpha < +\infty$, $\beta = +\infty$. If $\xi \in \partial D$, then $\operatorname{Im} \xi \leq -\alpha$ and from Lemma 3.3.1

$$\left[\left(1 - \frac{it}{\alpha}\right)\left(1 - \frac{t}{\xi}\right)\left(1 - \frac{t}{\bar{\xi}}\right)\right]^{-1}$$

is a characteristic function. This implies that each point of D has the frontier property† for the set $\mathscr{F}_D$ of all the Fourier–Stieltjes transforms of finite measures which are regular in D. Since $\mathscr{F}_D$ is closed for multiplication by a positive constant, addition, raising to a positive integer power and passing to the limit uniformly for each compact set of D, the Cartan–Thullen theorem implies the existence of an element $\hat{\mu}$ of $\mathscr{F}_D$ having D for domain of analyticity. Then $\hat{\mu}/\|\mu\|$ is a characteristic function having D for domain of anyalyticity.

In the general case $\alpha < +\infty$, $\beta < +\infty$, it follows from above that there exists a characteristic function $\hat{p}_1$ having $D \cup \{z \in C : \operatorname{Im} z > 0\}$ for domain of analyticity. In the same way, there exists a characteristic function $\hat{p}_2$ having $D \cup \{z \in C : \operatorname{Im} z < 0\}$ for domain of analyticity. Then $\frac{1}{2}(\hat{p}_1 + \hat{p}_2)$ is a characteristic function having D for domain of analyticity.

Theorem 3.3.6 If $\Gamma \subset R^n$ is an open convex set containing the origin, there exists an n-variate characteristic function having $R^n + i\Gamma$ for domain of analyticity.

Proof If Λ is an arbitrary convex set of R^n, there exists a positively homogeneous function h defined on R^n with values in $\bar{R}$ [that is, $h(\lambda\theta) = \lambda h(\theta)$ for any $\theta \in R^n$, $\lambda \in 0$] defined by‡

$$h(\theta) = \sup\{(x, \theta), x \in \Lambda\}.$$

The function h is the support function of Λ. If Λ is closed, then

$$\Lambda = \bigcap_{\theta \in R^n} \Pi_\theta$$

† For the notion of frontier property and for the Cartan–Thullen theorem, see Appendix A.4.
‡ For the results on convex sets used here, see for example, Choquet (1969, Section 21).

where $\Pi_\theta = \{x \in R^n : (x, \theta) \leq h(\theta)\}$ and Λ is uniquely determined by h.

If $\Gamma = R^n$, the theorem is clear. We suppose now that $\Gamma \neq R^n$, that is, there exists some $\theta \in R^n$ such that $h(\theta)$ is finite, h being the support function of Γ. From the preceding properties, it follows that

$$\bar{D} = \bigcap_{\theta \in R^n} P_\theta$$

where $D = R^n + i\Gamma$ and $P_\theta = \{z \in C^n : (\text{Im } z, \theta) \leq h(\theta)\}$. Each point $\xi \in \partial D$ belongs to some P_θ and we consider the characteristic function $\hat{q}$ defined by

$$\hat{q}(t) = \hat{p}(\text{pr}_\theta\, t/h(\theta)),$$

the function $\hat{p}$ being a univariate characteristic function having the half-space $\{z \in C : \text{Im } z \leq 1\}$ for domain of analyticity (its existence follows from the preceding theorem), $\hat{q}$ has $\{z \in C^n : (\text{Im } z, \theta) \leq h(\theta)\}$ for domain of analyticity and ξ has the frontier property for the set $\mathscr{F}_D$ of all the Fourier–Stieltjes transforms of finite measures which are regular in D. Since $\mathscr{F}_D$ is closed for the suitable operations, the Cartan–Thullen theorem implies the existence of an n-variate characteristic function having D for domain of analyticity.

It must be noted that Theorem 3.3.5 solves completely in the case $n = 1$ the problem of the description of the class of domains of C^n which are domains of analyticity of some characteristic functions, while this problem is not yet solved in the case $n > 1$.

3.4 SOME CHARACTERIZATION THEOREMS

We give now some simple characterizations of the most important probabilities. The simplest of these results is the easily proved.

Theorem 3.4.1 $p \in \mathscr{P}_n$ is degenerate if and only if its projections on n linearly independent vectors are degenerate.

Theorem 3.4.2 $p \in \mathscr{P}_n$ is a normal probability if and only if all its univariate projections are normal probabilities.

Proof The necessity is evident. To prove the sufficiency, we note that if, for some one-dimensional subspace E, $\text{pr}_E\, p$ is a normal probability, then $\text{pr}_E\, \hat{p}$ is an entire function of order 2 and without zeros. If this is true for any one-dimensional subspace, then $\hat{p}$ is an entire function of order 2 and without zeros. Therefore $\hat{p}$ is the exponential of some polynomial of degree 2, and from

$$\hat{p}(-t) = \overline{\hat{p}(t)}$$

it follows easily that p is a normal probability.

With the same method, we deduce from Marcinkiewicz's theorem [see Lukacs (1970, Corollary to Theorem 7.3.3)].

Theorem 3.4.3 If a characteristic function $\hat{p}$ admits for any $t \in R^n$ the representation

$$\hat{p}(t) = \exp(Q(t))$$

where Q is a polynomial, then p is a normal probability (perhaps degenerate).

Theorem 3.4.4 $p \in \mathscr{P}_n$ is an n-variate Poisson probability if and only if the two following conditions are satisfied:

(a) $\hat{p}$ is an entire characteristic function;
(b) all the margins of the probability p_y defined by (3.3.9) are univariate Poisson probabilities for any $y \in R^n$.

Proof The necessity is evident and we prove the sufficiency by induction on n. For any $t_1 \in C$ and $(y_2, \ldots, y_n) \in R^{n-1}$, we obtain from (b)

$$\hat{p}(t_1, iy_2, \ldots, iy_n) = \exp[\lambda_1(y_2, \ldots, y_n) + \lambda_2(y_2, \ldots, y_n)e^{it_1}] \tag{3.4.1}$$

where λ_j is a function with real values ($j = 1, 2$). From induction hypothesis, we obtain for any $y_1 \in R$ and $(t_2, \ldots, t_n) \in C^{n-1}$

$$\hat{p}(iy_1, t_2, \ldots, t_n) = \exp\left[\mu(y_1) + \sum_{\varepsilon:\varepsilon_1=0} v_\varepsilon(y_1)e^{i(\varepsilon, t)}\right] \tag{3.4.2}$$

where μ and v_ε are functions with real values. Comparing (3.4.1) and (3.4.2), we obtain the equation

$$\lambda_1(y_2, \ldots, y_n) + \lambda_2(y_2, \ldots, y_n)e^{-y_1} = \mu(y_1) + \sum_{\varepsilon:\varepsilon_1=0} v_\varepsilon(y_1)e^{-(\varepsilon, y)} + 2i\pi b(y)$$

where b is a continuous function with integer values. Since $b(0) = 0$, then $b(y) \equiv 0$. Since 1 and $\exp(-y_1)$ are linearly independent functions, we obtain

$$\lambda_1(y_2, \ldots, y_n) = l_1 + \sum_{\varepsilon:\varepsilon_1=0} m_{1,\varepsilon}e^{-(\varepsilon, y)},$$

$$\lambda_2(y_2, \ldots, y_n) = l_2 + \sum_{\varepsilon:\varepsilon_1=0} m_{2,\varepsilon}e^{-(\varepsilon, y)},$$

where l_j and $m_{j,\varepsilon}$ are real constants ($j = 1, 2$). This implies with the condition $\log \hat{p}(0) = 0$ that $\hat{p}$ admits the representation

$$\hat{p}(iy) = \exp\left[\sum_{\varepsilon} l_\varepsilon(e^{-(\varepsilon, y)} - 1)\right] \tag{3.4.3}$$

for any $y \in R^n$ where the l_ε are $2^n - 1$ real constants. Since $\hat{p}$ is an entire function and since the right side of (3.4.3) is also an entire function, we obtain

$$\hat{p}(t) = \exp\left[\sum_{\varepsilon} l_\varepsilon (e^{i(\varepsilon, t)} - 1)\right]$$

for any $t \in C^n$ and from Corollary 3.3.1, it follows easily that the l_ε are nonnegative, so that $\hat{p}$ is a Poisson probability.

Theorem 3.4.5 $p \in \mathscr{P}_n$ is the convolution of a normal probability and a Poisson probability if and only if the following conditions are satisfied:

(a) $\hat{p}$ is an entire characteristic function;

(b) all the margins of the probability p_y defined by (3.3.9) are convolutions of an univariate normal probability and an univariate Poisson probability for any $y \in R^n$.

Proof With the method used for the preceding theorem, we obtain

$$\hat{p}(t_1, iy_2, \ldots, iy_n) = \exp[\lambda_1(y_2, \ldots, y_n) + i\lambda_2(y_2, \ldots, y_n)t_1 - \lambda_3(y_2, \ldots, y_n)t_1^2 + \lambda_4(y_2, \ldots, y_n)e^{it_1}]$$

for any $t_1 \in C$, $(y_2, \ldots, y_n) \in R^{n-1}$ where λ_j is a function with real values $(j = 1, 2, 3, 4)$ and

$$\hat{p}(iy_1, t_2, \ldots, t_n) = \exp\left[\alpha(y_1) + i\sum_{j=2}^{n} \beta_j(y_1)t_j - \sum_{2 \le j \le k \le n} \gamma_{j,k} t_j t_k + \sum_{\varepsilon:\varepsilon_1 = 0} \delta_\varepsilon(y_1) e^{i(\varepsilon, t)}\right]$$

for any $y_1 \in R$, $(t_2, \ldots, t_n) \in C^{n-1}$ where α, β_j, $\gamma_{j,k}$, and δ_ε are functions with real values. From these relations, we obtain the equation

$$\lambda_1(y_2, \ldots, y_n) - \lambda_2(y_2, \ldots, y_n)y_1 + \lambda_3(y_2, \ldots, y_n)y_1^2 + \lambda_4(y_2, \ldots, y_n)e^{-y_1}$$
$$= \alpha(y_1) - \sum_{j=2}^{n} \beta_j(y_1)y_j + \sum_{2 \le j \le k \le n} \gamma_{j,k}(y_1)y_j y_k + \sum_{\varepsilon:\varepsilon_1 = 0} \delta_\varepsilon(y_1)e^{-(\varepsilon, y)}$$

for any $y = (y_1, \ldots, y_n) \in R^n$. From this we deduce that

$$\lambda_l(y_2, \ldots, y_n) = \alpha_l + \sum_{j=2}^{n} \beta_{l,j} y_j + \sum_{2 \le j \le k \le n} \gamma_{l,j,k} y_j y_k + \sum_{\varepsilon:\varepsilon_1 = 0} \delta_{l,\varepsilon} e^{-(\varepsilon, y)} \qquad (l = 1, 2, 3, 4)$$

where α_l, $\beta_{l,j}$, $\gamma_{l,j,k}$, and $\delta_{l,\varepsilon}$ are real constants and we obtain the representation

$$\hat{p}(iy) = \exp\Bigg[A + \sum_{j=1}^{n} B_j y_j + \sum_{1 \le j \le k \le n} C_{j,k} y_j y_k + \sum_{\varepsilon} D_\varepsilon e^{-(\varepsilon, y)} + \sum_{2 \le j \le k \le n} E_{j,k} y_1 y_j y_k + \sum_{j=2}^{n} F_j y_1^2 y_j + \sum_{2 \le j \le k \le n} G_{j,k} y_1^2 y_j y_k + \sum_{\varepsilon:\varepsilon_1 = 0} (H_\varepsilon y_1 + I_\varepsilon y_1^2) e^{-(\varepsilon, y)} + e^{-y_1}\Bigg(\sum_{j=2}^{n} J_j y_j + \sum_{2 \le j \le k \le n} K_{j,k} y_j y_k\Bigg)\Bigg]$$

valid for any $y \in R^n$ and, since $\hat{p}$ is an entire function, for any $y \in C^n$.

If we define ρ by

$$\rho(x, y) = \mathrm{Re}[\log \hat{p}(iy) - \log \hat{p}(x + iy)]$$

for any $x \in R^n$, $y \in R^n$, we obtain from Corollary 3.3.2 the relation

$$\rho(x, y) \ge 0 \tag{3.4.4}$$

for any x, y. But we easily obtain

$$\begin{aligned}\rho(x, y) = {} & \sum_{1 \le j \le k \le n} C_{j,k} x_j x_k + 2 \sum_{\varepsilon} D_\varepsilon e^{-(\varepsilon, y)} \sin^2(\tfrac{1}{2}(\varepsilon, x)) \\ & + \sum_{2 \le j \le k \le n} E_{j,k}(x_1 y_j y_k + x_j y_1 y_k + x_k y_1 y_j) \\ & + \sum_{j=2}^{n} F_j(y_j x_1^2 + 2x_1 y_1 x_j) \\ & + \sum_{2 \le j \le k \le n} G_{j,k}[x_1^2 y_j y_k + x_j x_k y_1^2 - x_1^2 x_j x_k \\ & + 2x_1 y_1(x_j y_k + x_k y_j)] \\ & + \sum_{\varepsilon:\varepsilon_1 = 0} e^{-(\varepsilon, y)}[H_\varepsilon(y_1 \sin^2(\tfrac{1}{2}(\varepsilon, x)) + x_1 \sin(\varepsilon, x)) \\ & + I_\varepsilon(y_1^2 \sin^2(\tfrac{1}{2}(\varepsilon, x)) + x_1^2 \cos(\varepsilon, x) + 2x_1 y_1 \sin(\varepsilon, x))] \\ & + e^{-y_1} \sum_{j=2}^{n} \Bigg[J_j(y_j \sin^2(\tfrac{1}{2}x_1) + x_j \sin x_1) \\ & + \sum_{2 \le j \le k \le n} K_{j,k}(y_j y_k \sin^2(\tfrac{1}{2}x_1) + x_j x_k \cos x_1) \\ & + (x_j y_k + x_k y_j) \sin x_1\Bigg].\end{aligned}$$

When $x_1 \to \infty$, we find

$$\rho(x, y) = x_1^2\Bigg[C_{1,1} + \sum_{j=2}^{n} F_j y_j + \sum_{2 \le j \le k \le n} G_{j,k}(y_j y_k - x_j x_k) + \sum_{\varepsilon:\varepsilon_1 = 0} I_\varepsilon e^{-(\varepsilon, y)} \cos(\varepsilon, x)\Bigg] + O(x_1)$$

and from (3.4.4), we deduce that the expression within brackets in the right side of the preceding relation is always nonnegative and this implies easily that

$$F_j = G_{j,k} = I_\varepsilon = 0.$$

When $x_j \to \infty$ $(j \neq 1)$, we find

$$\rho(x, y) = x_j^2[C_{j,j} + K_{j,j} e^{-y_1} \cos x_1] + O(x_j)$$

and this implies

$$K_{j,j} = 0.$$

When $x_j = x_k \to \infty$, we find

$$\rho(x, y) = x_j^2[C_{j,j} + C_{j,k} + C_{k,k} + K_{j,k} e^{-y_1} \cos x_1] + O(x_j)$$

and this implies

$$K_{j,k} = 0 \qquad (j \neq k).$$

When $y_1 \to \infty$, we find

$$\rho(x, y) = y_1\Bigg[\sum_{2 \le j \le k \le n} E_{j,k}(x_j y_k + x_k y_j) + \sum_{\varepsilon:\varepsilon_1 = 0} H_\varepsilon e^{-(\varepsilon, y)} \sin^2(\tfrac{1}{2}(\varepsilon, x))\Bigg] + O(1)$$

and this implies with (3.4.4) that

$$E_{j,k} = 0.$$

When $y_1 \to -\infty$, we find

$$\rho(x, y) = e^{-y_1}\Bigg[2 \sum_{\varepsilon:\varepsilon_1 = 0} D_\varepsilon e^{-(\varepsilon, y) + y_1} \sin^2(\tfrac{1}{2}(\varepsilon, x)) + \sum_{j=2}^{n} J_j(y_j \sin^2(\tfrac{1}{2}x_1) + x_j \sin x_1)\Bigg] + O(1)$$

so that

$$J_j = 0 \qquad \text{and} \qquad D_\varepsilon \ge 0 \qquad \text{if} \quad \varepsilon_1 = 1.$$

In the same way, $y_j \to -\infty$ $(j \neq 1)$ implies

$$H_\varepsilon = 0 \qquad \text{and} \qquad D_\varepsilon \geq 0 \qquad \text{if} \quad \varepsilon_j = 1.$$

Now, if

$$Q(x) = \sum_{1 \leq j \leq k \leq n} C_{j,k} x_j x_k$$

is negative for some x_0, then

$$\lim_{r \to \infty} \rho(rx_0, y) = -\infty,$$

so that Q is a nonnegative quadratic form. Finally, from $\hat{p}(0) = 1$, we obtain

$$A = -\sum_\varepsilon D_\varepsilon$$

and we have

$$\hat{p}(t) = \exp\left[i(b, t) - Q(t) + \sum_\varepsilon D_\varepsilon(e^{i(\varepsilon, t)} - 1)\right]$$

where $b = (B_1, \ldots, B_n) \in R^n$, so that p is the convolution of a normal and a Poisson probability.

3.5 AN EXTENSION OF THE NOTION OF ANALYTIC CHARACTERISTIC FUNCTIONS

We introduce the following definition. If Γ is a subset of R^n, an n-variate probability p belongs to $\mathscr{A}_\Gamma$ if

$$\int e^{-(y, u)} p(du)$$

converges for any $y \in \Gamma$. Then the integral

$$\int e^{i(t, u)} p(du)$$

exists for any $t \in R^n + i\Gamma$. We can suppose (without loss of generality) that Γ is convex and contains the origin.

This definition extends to the case of n variables the notion of boundary characteristic function [see Lukacs (1970, Chapter 11)]. When the origin is an interior point of Γ, it follows from Theorem 3.3.2 that $\mathscr{A}_\Gamma$ is the class of all the probabilities having characteristic functions analytic in $R^n + i\Gamma$.

The class $\mathscr{A}_\Gamma$ is characterized by the following result.

Theorem 3.5.1 If Γ is a convex set containing the origin, the following conditions are equivalent:

(a) p belongs to $\mathscr{A}_\Gamma$;
(b) $j(\theta) \leq -\limsup_{u\to\infty}(1/u)\log \hat{p}(\{x:(x,\theta)\leq -u\})$ for any $\theta \in R^n$ where j is the gauge of Γ [if for some $u > 0$, $p(\{x:(x,\theta)\leq -u\}) = 0$, then the limit in the second member is by convention $-\infty$];
(c) for any $y \in \Gamma\backslash\{0\}$, there exists a function f_y of the variable k which is continuous in $\{k: 0 \leq \operatorname{Im} k \leq 1\}$, regular in $\{k: 0 < \operatorname{Im} k < 1\}$, and satisfies

$$f_y(k) = \hat{p}(ky)$$

for any $k \in R$;
(d) for any $y \in \Gamma\backslash\{0\}$ and any ϕ orthogonal to y, there exists a function $g_{\phi,y}$ of the variable k which is continuous in $\{k: 0 \leq \operatorname{Im} k \leq 1\}$, regular in $\{k: 0 < \operatorname{Im} k < 1\}$, and satisfies

$$g_{\phi,y}(k) = \hat{p}(ky + \phi)$$

for any $k \in R$.

Moreover

$$g_{0,y} = f_y \tag{3.5.1}$$

and

$$g_{\phi,y}(k) = \int e^{i(ky+\phi,u)}p(du). \tag{3.5.2}$$

Proof If we define $g_{\phi,y}$ by (3.5.2), we find easily that (a) implies (d) and it is evident that (3.5.1) and (d) imply (c). Now, if we note that (c) can be stated as $\mathrm{pr}_y\,\hat{p}$ is the boundary of a regular function, it follows from the univariate case [see Lukacs (1970, Theorem 11.1.1 and Corollary 5 to Theorem 11.1.1)] that (c) implies (a) and is equivalent to (b).

In the hypothesis of Theorem 3.5.1, the integral $\int e^{i(t,u)}p(du)$ which exists for any $t \in R^n + i\Gamma$ will be denoted by $\hat{p}(t)$. For example, (3.5.2) is equivalent to

$$g_{\phi,y}(k) = \hat{p}(ky + \phi) \tag{3.5.3}$$

for any $k \in C$ satisfying $0 < \operatorname{Im} k < 1$.

From this theorem, we can deduce that a probability belonging to $\mathscr{A}_\Gamma$ has most of the properties of a probability with an analytic characteristic function. As examples, we have the following corollaries.

Corollary 3.5.1 (*Ridge property*) If $p \in \mathscr{A}_\Gamma$, then

$$|\hat{p}(x + iy)| \leq \hat{p}(iy)$$

for any $x \in R^n$, $y \in \Gamma$. If $\hat{p}$ has no zeros in $R^n + i\Gamma$, then

$$\operatorname{Re}[\log \hat{p}(iy) - \log \hat{p}(x + iy)] \geq 0$$

for any $x \in R^n$, $y \in \Gamma$.

Proof This follows immediately from (3.5.3).

Corollary 3.5.2 If $p \in \mathscr{A}_\Gamma$, then the function f_y defined by

$$f_y(t) = \hat{p}(t + iy)/\hat{p}(iy)$$

is a characteristic function for any $y \in \Gamma$.

Proof This is proved in the same way as Corollary 3.3.5.

Corollary 3.5.3 If $p \in \mathscr{A}_\Gamma$, there exist some $C > 0$ and some $m \in R^n$ such that

$$\hat{p}(iy) \geq C \exp(-(m, |y|)) \tag{3.5.4}$$

for any $y \in \Gamma$.

Proof Let $a_1, \ldots, a_{2^n} \in R^n$ $[a_j = (a_{j,1}, \ldots, a_{j,n})]$ defined by

$$p(\{x \in R^n : x_1 \leq a_{1,1}, x_2 \leq a_{1,2}, \ldots, x_n \leq a_{1,n}\}) = C_1 > 0,$$

$$p(\{x \in R^n : x_1 \geq a_{2,1}, x_2 \leq a_{2,2}, \ldots, x_n \leq a_{2,n}\}) = C_2 > 0,$$

$$\vdots$$

$$p(\{x \in R^n : x_1 \geq a_{2^n,1}, x_2 \geq a_{2^n,2}, \ldots, x_n \geq a_{2^n,n}\}) = C_{2^n} > 0.$$

Then if

$$C = \inf_{1 \leq j \leq 2^n} C_j, \qquad m = (m_1, \ldots, m_n), \qquad m_k = \sup_{1 \leq j \leq 2^n} |a_{j,k}|,$$

from

$$\hat{p}(iy) = \int e^{-(y,u)} p(du),$$

we obtain easily (3.5.4).

For any convex set Γ containing the origin, we have also the existence of probabilities which belong to $\mathscr{A}_\Gamma$, but do not belong to $\mathscr{A}_{\Gamma'}$ for any convex set Γ' greater than Γ.

3.6 CONVEX SUPPORT OF SIGNED MEASURES

We study now the convex support of a signed measure μ, that is, the smallest closed convex set on which μ is concentrated (or the convex hull of the support of μ). The simplest of these convex supports are the whole space of R^n (in this case, there is nothing to say) and a half-space $\{x \in R^n : (x, \theta) \leq m\}$ where $\theta \in R^n \backslash \{0\}$ and $m \in R$. For this last case, we introduce the following definition.

If $\theta \in R^n \backslash \{0\}$, a signed measure μ is bounded in the direction θ if there exists some $m \in R^n$ such that μ is concentrated on $\{x \in R^n : (x, \theta) \leq m\}$. We define the *extremity* of μ in the direction θ (and we denote it by $\text{ext}_\theta\, \mu$) as the infimum of the m having this property.†

We have then

Theorem 3.6.1 $\mu \in \mathscr{M}_n$ is bounded in the direction θ if and only if for any $\phi \in R^n$ orthogonal to θ, there exists a function f_ϕ of one complex variable such that:

(a) f_ϕ is continuous in the half-plane $\{k \in C : \text{Im } k \leq 0\}$ and regular and of exponential type (independent of ϕ) in the half-plane $\{k \in C : \text{Im } k < 0\}$;
(b) $f_\phi(k) = \hat{\mu}(k\theta + \phi)$ for $k \in R$, $\phi \perp \theta$.

Then

$$\text{ext}_\theta\, \mu = \sup_{\phi \perp \theta} \limsup_{l \to \infty} \frac{\log |f_\phi(-il)|}{l}. \tag{3.6.1}$$

Proof *Necessity* If μ is bounded in the direction θ and if we define f_ϕ by

$$f_\phi(k) = \int e^{i(k\theta + \phi, u)} \mu(du),$$

then it is easily seen that these functions have all the required properties. Moreover

$$|f_\phi(-il)| \leq \|\mu\| e^{l \text{ext}, \mu},$$

and from this we have

$$\kappa \leq \text{ext}_\theta\, \mu \tag{3.6.2}$$

where κ is the right side of (3.6.1).

Sufficiency We can suppose without loss of generality that $\theta = e_1$ and we consider $a = (a_1, \ldots, a_n) \in R^n$, $b = (b_1, \ldots, b_n) \in R^n$, $h < a_1$, $a < b$. If $\{x \in R^n : a < x < b\}$ is a continuity set of μ, then, from Theorem 2.3.1,

$$(2\pi)^n \mu(\{x \in R^n : a < x < b\})$$

$$= \lim_{T_2 \to \infty} \cdots \lim_{T_n \to \infty} \int_{C_T} \left(\prod_{j=2}^{n} \frac{e^{-iu_j a_j} - e^{-iu_j b_j}}{iu_j} \right) H(u) \lambda(du)$$

† It must be noted that $\text{ext}_\theta\, \mu = h(\theta)$ where h is the support function of the convex support of μ.

where $u = (u_2, \ldots, u_n)$, $T = (T_2, \ldots, T_n)$, and $C_T = \{x \in R^{n-1} : -T \leq x \leq T\}$. Then H is defined by

$$H(u) = \lim_{r \to \infty} \int_{-r}^{r} \frac{e^{-ika_1} - e^{-ikb_1}}{ik} f_u(k)\, dk.$$

If $\varepsilon = \frac{1}{2}(a_1 - h)$ and if we define g_u by

$$g_u(k) = f_u(k) e^{-i(h+\varepsilon)k}(1 - e^{i(b_1 - a_1)k}),$$

we obtain a function which is continuous in $\{k \in C : \text{Im } k \leq 0\}$, regular and of exponential type in $\{k \in C : \text{Im } k < 0\}$, and bounded on the real axis and on $\{k \in C : \text{Re } k = 0, \text{Im } k \leq 0\}$. From the Phragmén–Lindelöf theorem (see Appendix A.4), we deduce the existence of a constant C such that $|g_u(k)| \leq C$ for any k satisfying $\text{Im } k \leq 0$. Since

$$H(u) = \lim_{r \to \infty} \int_{-r}^{r} \frac{g_u(k) e^{-i\varepsilon k}}{ik} dk,$$

if we replace the integration segment $[-r, r]$ by the half-circle with center at the origin and radius r located in the lower half-plane, we obtain

$$|H(u)| \leq \lim_{r \to \infty} \int_0^{\pi} g_u(k) e^{-\varepsilon r \sin k}\, dk = 0.$$

It follows then that

$$\mu(\{x : a < x < b\}) = 0$$

if $(a, e_1) \geq \kappa$, so that μ is bounded in the direction θ and

$$\text{ext}_\theta\, \mu \geq \kappa. \tag{3.6.3}$$

Equation (3.6.1) follows immediately from (3.6.2) and (3.6.3).

Since an arbitrary closed convex set Γ is an intersection of half-planes of the preceding type, the preceding theorem solves also the general case. In particular, we have

Corollary 3.6.1 $\mu \in \mathcal{M}_n$ is concentrated on a compact set if and only if there exists an entire function f of exponential type such that

$$\hat{\mu}(t) = f(t)$$

for any $t \in R^n$. Moreover, if h is the support function of the convex support of μ, then

$$h(\theta) = \sup_{\phi \perp \theta} \limsup_{l \to \infty} \frac{\log |f(\phi - il\theta)|}{l}$$

for any $\theta \in R^n$.

For the following result, we will use these notations: $L^2(R^n)$ is the set of the functions f defined on R^n with complex values and satisfying

$$\int |f(x)|^2\lambda(dx) < +\infty;$$

if $f \in L^2(R^n)$, we define $\|f\|_2$ by

$$\|f\|_2 = \left(\int |f(x)|\,\lambda(dx)\right)^{1/2};$$

and if $f \in L^2(R^n)$, $g_u \in L^2(R^n)$, we say that g_u *converges in mean to* f when u tends to u_0 and we write

$$f = \underset{u \to u_0}{\text{l.i.m.}}\ g(u)$$

if

$$\lim_{u \to u_0} \|f - g_u\|_2 = 0.$$

We have then the following result which is closely related to Corollary 3.6.1 (the case $n = 1$ is known as the *Paley–Wiener theorem*).

Theorem 3.6.2 The following conditions are equivalent:

(a) f is an entire function of exponential type belonging to $L^2(R^n)$;

(b) f is the Fourier–Stieltjes transform of an absolutely continuous finite signed measure μ with a Radon–Nikodym derivative $\phi \in L^2(R^n)$ and concentrated on a compact set Γ.

Moreover, if h is the support function of the convex support of μ, then

$$h(\theta) = \sup_{\phi \perp \theta} \limsup_{l \to \infty} \frac{\log |f(\phi - il\theta)|}{l} \tag{3.6.4}$$

for any $\theta \in R^n$.

Proof We will use the following result on Fourier integrals.† If $f \in L^2(R^n)$, there exists a function $\phi \in L^2(R^n)$ such that

$$f(t) = \underset{u_1 \to \infty, \ldots, u_n \to \infty}{\text{l.i.m.}} \int_{-u_1}^{u_1} \cdots \int_{-u_n}^{u_n} e^{i(t, x)}\phi(x)\lambda(dx) \tag{3.6.5}$$

† A particular case of this result is given by Bochner and Chandrasekharan (1949, p. 120); a modification of their proof gives our statement.

and ϕ satisfies

$$(2\pi)^n \phi(x) = \underset{v_1 \to \infty, \ldots, v_n \to \infty}{\text{l.i.m.}} \int_{-v_1}^{v_1} \cdots \int_{-v_n}^{v_n} e^{-i(t,x)} f(t) \lambda(dt). \tag{3.6.6}$$

From this result and Corollary 3.6.1, we deduce that (b) implies (a). Conversely, if $f \in L^2(R^n)$, we obtain the existence of a function $\phi \in L^2(R^n)$ defined by (3.6.6) and satisfying (3.6.5). If f is an entire function of exponential type, we deduce exactly as in Theorem 3.6.1 that

$$\phi(x) = 0$$

if $(x, \theta) \geq \kappa$ where κ is the right side of (3.6.4). Therefore ϕ is absolutely integrable and from Corollary 2.3.1, f is the Fourier–Stieltjes transform of the signed measure μ having ϕ for its Radon–Nikodym derivative. The last assertion follows directly from Corollary 3.6.1.

For probabilities, we can find simpler conditions than those of the preceding results.

Theorem 3.6.3 $p \in \mathscr{P}_n$ is bounded in the direction θ if and only if the two following conditions are satisfied:

(a) $p \in \mathscr{A}_\Gamma$ where $\Gamma = \{x \in R^n : x = -\lambda\theta, \lambda \geq 0\}$;
(b) there exists some real constant C such that

$$\hat{p}(-iy\theta) \leq e^{Cy}$$

for any $y > 0$.

We have then

$$\text{ext}_\theta\, p = \lim_{y \to \infty} \frac{\log \hat{p}(-iy\theta)}{y}.$$

Proof From Theorems 3.6.1 and 3.5.1, the first part of the theorem follows immediately. Since

$$\left| \int e^{i(\theta y + \phi, u)} p(du) \right| \leq \hat{p}(-i\theta \operatorname{Im} y)$$

for any $\phi \in R^n$ orthogonal to θ, from (3.6.1) we have

$$\text{ext}_\theta\, p = \limsup_{y \to \infty} \frac{\log \hat{p}(-iy\theta)}{y}.$$

The existence of $\lim(\log \hat{p}(-iy\theta)y^{-1})$ follows easily from the convexity of $\log \hat{p}(-iy\theta)$ [see Lukacs (1970, p. 202)].

In the same way, we have from Corollary 3.6.1

Theorem 3.6.4 $p \in \mathscr{P}_n$ is concentrated on a compact set if and only if $\hat{p}$ is an entire function of exponential type. If h is the support function of the convex support of p, then

$$h(\theta) = \lim_{y \to \infty} \frac{\log \hat{p}(-iy\theta)}{y}$$

for any $\theta \in R^n$.

Finally, we prove the following extension of Corollary 3.6.1 which will be useful in the following.

Theorem 3.6.5 Let f be the function defined on R^n by the formula

$$f(t) = \int \left(e^{i(t,u)} - 1 - \frac{i(t, u)}{1 + \|u\|^2}\right) \frac{1 + \|u\|^2}{\|u\|^2} \mu(du)$$

where μ is a finite signed measure having no atom at the origin. If μ has a compact support, then f can be continued in the whole complex plane as an entire function of exponential type.

Proof The case $n = 1$ is a direct consequence of Lemma 9.1.1 of Lukacs (1970) and the general case follows from this one by projection.

NOTES

Theorems 3.1.1 and 3.1.2, respectively, are due to Pólya (1949) and to Dugué and Girault (1955).

The study of the link between moments of even order and derivatives of characteristic functions is given by Wintner (1938). The extension to derivatives of odd order is due to Zygmund (1947), Pitman (1956), and Morucci (1966).

The first study of univariate analytic characteristic functions is due to Raikov (1938b) who proved Theorem 3.3.1 for a strip symmetric with respect to the real axis. Theorem 3.3.1 in its full generality (always in the case $n = 1$) is due to Lukacs (1957), while Dugué (1951a) has studied the consequences of the ridge property.

In the case $n > 1$, Theorem 3.3.1 has been proved independently by Cuppens (1967) and Ostrovskiĭ (1966a). The study of the domains which are domains of analyticity of characteristic functions is due to Ostrovskiĭ (1966a).

The characterization of the n-variate normal probabilities by its projections is due to Cramér (1936), while the one of the n-variate Poisson probability has been given by Teicher (1954).

The class $\mathscr{A}_\Gamma$ in the case $n = 1$ was first studied by Marcinkiewicz (1938) and later by Esseen (1965) and Ramachandran (1966). In the case $n > 1$, this notion is due to Cuppens (1967).

Theorem 3.6.2 is due in the case $n = 1$ to Paley and Wiener (1934) and in the general case to Plancherel and Pólya (1937). With the same method, Pólya (1949) proved Theorem 3.6.3 in the case $n = 1$, the general case being due to Cuppens (1967).

Chapter 4

Decomposition Theorems

4.1 INDECOMPOSABLE PROBABILITIES

If p, p_1, and p_2 are n-variate probabilities satisfying

$$p = p_1 * p_2, \tag{4.1.1}$$

we say that p_j is a *factor* (or *divides*) p ($j = 1, 2$) and that (4.1.1) is a decomposition of p. If p is a given probability, we can raise the problem of finding all its factors. As we will see in the following, this problem is a very difficult one for arbitrary probabilities.

Evidently, we have for any $\alpha \in R^n$ a decomposition (4.1.1) with $p_1 = p * \delta_\alpha$ and $p_2 = \delta_{-\alpha}$. We call such a decomposition a *trivial* one and introduce the following definition. A probability which has only trivial decompositions is *indecomposable* (otherwise, it is decomposable). For convenience, a degenerate probability is not indecomposable.

There is no general method for determining if a given probability is indecomposable. Nevertheless, we can make the following remark. If (4.1.1) is satisfied, then, from Corollary 2.5.2, we have

$$S(p) = \overline{S(p_1) + S(p_2)}. \tag{4.1.2}$$

If $S(p)$ is compact, then $S(p_j)$ is also compact ($j = 1, 2$) and (4.1.2) becomes

$$S(p) = S(p_1) + S(p_2). \tag{4.1.3}$$

Therefore, if p is decomposable, that implies that the set $S(p)$ can be decomposed according to formula (4.1.2) with sets $S(p_j)$ having at least two points ($j = 1, 2$). From this remark, it follows that the fact that a purely atomic probability is decomposable is rather scarce. We can make this statement precise with

Theorem 4.1.1 An n-variate probability p which is concentrated on an independent set is indecomposable.†

Proof This follows immediately from the preceding remark and the fact that if A_1 and A_2 have at least two points, then $A_1 + A_2$ is not an independent set.

Corollary 4.1.1 The set of all the indecomposable purely atomic n-variate probabilities is dense (with respect to complete convergence) in the set of all the n-variate probabilities.

Proof In each cube

$$A_{j,\,k_1,\,\ldots,\,k_n} = \{x = (x_1, \ldots, x_n) : k_m j^{-1} \leq x_m < k_{m+1} j^{-1},\ m = 1, \ldots, n\}$$

($j = 1, 2, \ldots; k_m = 0, \pm 1, \pm 2, \ldots, \pm j^2$), we can choose a point $\lambda_{j,\,k_1,\,\ldots,\,k_n}$ so that $\Lambda_j = \{\lambda_{j,\,k_1,\,\ldots,\,k_n}\}$ is an independent set for each j. If p is an arbitrary probability, we define for any j such that

$$c_j = p\left(\bigcup_{k_1,\,\ldots,\,k_n} A_{j,\,k_1,\,\ldots,\,k_n}\right) > 0$$

a probability p_j by

$$p_j(B) = c_j^{-1} \sum_{\lambda_{j,k_1,\ldots,k_n} \in B} p(A_{j,\,k_1,\,\ldots,\,k_n})$$

for any $B \in \mathscr{B}_n$. Then p_j is concentrated on Λ_j and

$$\lim_{j \to \infty} p_j \overset{c}{=} p.$$

Theorem 4.1.1 implies then the corollary.

We prove now

Theorem 4.1.2 Let $S_{m,\,r}$ be a sphere in R^3:

$$S_{m,\,r} = \{x \in R^3 : \|x - m\| = r\}$$

($m \in R^3, r > 0$). If p is a probability concentrated on $S_{m,\,r}$, then p is indecomposable.

† For a definition, see Appendix A.3.

Proof From (4.1.3), it suffices to prove that $S_{m,r}$ cannot be decomposed as $S_{m,r} = A + B$ where A and B have at least two points. Without loss of generality, we can suppose that $m = re_1$. Then $0 \in S_{m,r}$ and $S_{m,r} \subset \{x = (x_1, x_2, x_3) : x_1 \geq 0\}$. It follows easily that

$$A \subset \{x = (x_1, x_2, x_3) : x_1 \geq a\}, \qquad B \subset \{x = (x_1, x_2, x_3) : x_1 \geq b\}$$

with $a + b = 0$, so that we can suppose that $a = b = 0$. Then

$$0 \in A \subset S_{m,r}, \qquad 0 \in B \subset S_{m,r}.$$

If $\alpha \in A$ and $\beta \in B$, then the points 0, α, β, $\alpha + \beta$ belong to $S_{m,r}$ and are the summits of a rectangle, so that $\alpha \perp \beta$. If A has two points 0 and α, then B is contained in the circle $S_{m,r} \cap \alpha^{\perp}$ and $A + B \neq S_{m,r}$. If A has more than two points, then B has two points and again $A + B \neq S_{m,r}$ and the theorem is proved.

This theorem has an analogy in the case $n = 2$. If we consider the uniform probability p on $S_{0,r}$, then for any $\theta \in R^3$, $\mathrm{pr}_\theta\, p$ is a uniform measure on the segment $\{x : x = l\theta, \ -r \leq l \leq r\}$ [cf. Feller (1966, p.497)] and $\mathrm{pr}_\theta\, p$ is decomposable [the factors of a uniform probability on a segment have been described by Lewis (1967) and Tortrat (1969)].

Therefore, an indecomposable probability can have decomposable projections. Also, a decomposable probability can have indecomposable projections. For example, the bivariate probability p having for characteristic function

$$\hat{p}(t) = \tfrac{1}{4}(1 + e^{it_1})(1 + e^{it_2}) \tag{4.1.4}$$

is decomposable, but $\mathrm{pr}_{e_1}\, p$ and $\mathrm{pr}_{e_2}\, p$ are indecomposable. Nevertheless, a decomposable probability cannot have "too much" indecomposable projections as it is shown by

Theorem 4.1.3 Let $p \in \mathscr{P}_n$ and $\theta_1, \ldots, \theta_m$ be $m \geq 2n - 1$ vectors of R^n such that n of these vectors are linearly independent. If $\mathrm{pr}_{\theta_j}\, p$ is indecomposable $(j = 1, \ldots, m)$, then p is indecomposable.

Proof If $p = q * r$ $(q \in \mathscr{P}_n, r \in \mathscr{P}_n)$, then $\mathrm{pr}_{\theta_j}\, q$ and $\mathrm{pr}_{\theta_j}\, r$ divide $\mathrm{pr}_{\theta_j}\, p$ and, by assumption, for any j, one of the $\mathrm{pr}_{\theta_j}\, q$ or $\mathrm{pr}_{\theta_j}\, r$ is degenerate. Use the fact that $m \geq 2n - 1$. Then one of the q or r has the property that its projections on n linearly independent vectors are degenerate and, from Theorem 3.4.1, is degenerate. Therefore, p is indecomposable.

Equation (4.1.4) shows that the condition $m \geq 2n - 1$ cannot be weakened.

For other results on indecomposable probabilities, see Parthasarathy (1967), and for examples of univariate indecomposable probabilities, see Lukacs (1970, pp. 180–189).

4.2 INFINITELY DIVISIBLE PROBABILITIES

We introduce now another class of probabilities which is important for decomposition theory.

An n-variate probability p is *infinitely divisible if*, for any positive integer k, there exists some probability p_k such that

$$p = p_k^{*k}. \tag{4.2.1}$$

We denote by I^n the set of all the n-variate infinitely divisible probabilities. From the convolution theorem, (4.2.1) is equivalent to

$$\hat{p} = (\hat{p}_k)^k. \tag{4.2.2}$$

If $\hat{p}$ is the characteristic function of an infinitely divisible probability, we say also that $\hat{p}$ is infinitely divisible.

Example 1 Degenerate probabilities are infinitely divisible.

Example 2 Normal probabilities and Poisson-type probabilities are infinitely divisible. [This follows immediately from (4.2.2).]

Example 3 If μ is a finite measure, then $\exp \mu/\exp\|\mu\|$ is an infinitely divisible probability. [This follows from Corollary 2.9.2 and (4.2.1).]

Example 4 If $p \in I^n$ and if E is a subspace of R^n, then $\mathrm{pr}_E\, p$ is infinitely divisible. [This follows from Corollary 2.5.8 and (4.2.1).] We will prove in the following that the converse is false.

We study now the most important properties of these probabilities.

Theorem 4.2.1 If $p \in I^n$, then $\hat{p}$ has no real zeros.

Proof Since $\hat{p}(0) = 1$ and since $\hat{p}$ is continuous, then, if $\hat{p}$ has some zeros, there exists some $m \in R^n$ such that $\hat{p}(m) = 0$ and $\hat{p}(t) \neq 0$ for $\|t\| < \|m\|$. Then $\log \hat{p}(t)$ is defined for $\|t\| < \|m\|$ and if $\hat{p}_k$ is defined by (4.2.2), then

$$\hat{p}_k(t) = \exp((1/k) \log \hat{p}(t)) = (\hat{p}(t))^{1/k}$$

for $\|t\| < \|m\|$. If

$$g = \lim_{k \to \infty} \hat{p}_k\,,$$

then

$$g(t) = \begin{cases} 1 & \text{if } \hat{p}(t) \neq 0, \\ 0 & \text{if } \hat{p}(t) = 0. \end{cases}$$

Therefore $g(t) = 1$ for $\|t\| < \|m\|$ and, by the continuity theorem, g is a characteristic function and is continuous everywhere. This contradicts the fact that $g(m) = 0$ and proves the theorem.

Theorem 4.2.2 If q and r are infinitely divisible, then $p = q * r$ is infinitely divisible.

Proof This follows immediately from

$$\hat{p} = \hat{q}\hat{r} = (\hat{q}_k)^k(\hat{r}_k)^k = (\hat{q}_k\hat{r}_k)^k$$

and from Corollary 2.5.7.

Corollary 4.2.1 If $\hat{p}$ is infinitely divisible, then $|\hat{p}|$ is an infinitely divisible characteristic function.

Proof If

$$p = p_k^{*k},$$

it is clear that

$$\tilde{p} = \tilde{p}_k^{*k},$$

so that if p is infinitely divisible, then the symmetric $\tilde{p}$ of p is also infinitely divisible. $|\hat{p}|^2$ is the characteristic function of $p * \tilde{p}$ which is infinitely divisible by the preceding theorem. The corollary follows then from the definition (4.2.2).

Theorem 4.2.3 If $p_j \in I^n$ $(j = 1, 2, \ldots)$ and if $\{p_j\}$ converges completely to p, then $p \in I^n$.

Proof If $p_j \xrightarrow{c} p$, then, for any positive integer k,

$$\lim_{j\to\infty} |\hat{p}_j|^{2/k} = |\hat{p}|^{2/k}.$$

From Corollary 4.2.1 and the continuity theorem, it follows that $|\hat{p}|^{2/k}$ is a characteristic function, so that $|\hat{p}|^2$ is an infinitely divisible characteristic function. Therefore, from Theorem 4.2.1, $\hat{p}$ has no real zeros and $\log \hat{p}$ is defined on R^n. Since

$$\lim_{j\to\infty} (\hat{p}_j)^{1/k} = \lim_{j\to\infty} \exp((1/k)\log \hat{p}_j) = \exp((1/k)\log \hat{p}) = \hat{p}^{1/k},$$

it follows from the continuity theorem that $\hat{p}^{1/k}$ is a characteristic function for any positive integer k, so that $\hat{p}$ is infinitely divisible.

Corollary 4.2.2 If $\hat{p}$ is infinitely divisible, $\hat{p}^\alpha$ is an infinitely divisible characteristic function for any positive constant α.

Proof If $\alpha = 1/k$, this follows from the definition. From Theorem 4.2.2, this is true for any positive rational α and, from Theorem 4.2.3, we deduce easily the corollary.

Theorem 4.2.4 (*De Finetti's theorem*) $p \in I^n$ if and only if there exists a sequence of finite measures $\{\mu_j\}$ such that

$$\exp \mu_j / \exp\|\mu_j\| \xrightarrow{c} p. \tag{4.2.3}$$

Proof If $\mu_j = jp_j$ [where p_j is defined by (4.2.1)], from

$$\hat{p} = \lim_{j \to \infty} \exp(j(\hat{p}_j - 1)),$$

we obtain (4.2.3). The converse follows immediately from Example 3 and Theorem 4.2.3.

Theorem 4.2.5 $p \in I^n$ if and only if p is the complete limit of a sequence of finite convolutions of Poisson-type probabilities.

Proof The sufficiency follows from Example 2 and Theorems 4.2.2 and 4.2.3 while the necessity follows from De Finetti's theorem, Corollary 2.9.5, and the fact that any finite measure is the complete limit of a sequence of purely atomic finite measures with a finite number of atoms.

We introduce the following definition. A function f defined on R^n with complex values is a *negative-definite* function if

$$\sum_{j=1}^{m} \sum_{k=1}^{m} \xi_j \overline{\xi_k} (f(t_j - t_k) - f(t_j) - f(-t_k)) \geq 0 \tag{4.2.4}$$

for any positive integer m, any complex constants $\xi_1, \ldots, \xi_m$, and any $t_1, \ldots, t_m$ belonging to R^n.

Lemma 4.2.1 Let f be a function defined on R^n with complex values and satisfying $f(0) = 0$. Then f is a negative-definite function if and only if

$$\sum_{j=1}^{m} \sum_{k=1}^{m} \xi_j \overline{\xi_k} f(t_j - t_k) \geq 0 \tag{4.2.5}$$

for any positive integer m, any $t_1, \ldots, t_m$ belonging to R^n, and any complex constants $\xi_1, \ldots, \xi_m$ satisfying

$$\sum_{j=1}^{m} \xi_j = 0. \tag{4.2.6}$$

Proof *Necessity* If (4.2.6) is satisfied, then

$$\sum_{j=1}^{m} \sum_{k=1}^{m} \xi_j \overline{\xi_k} (f(t_j - t_k) - f(t_j) - f(-t_k)) = \sum_{j=1}^{m} \sum_{k=1}^{m} \xi_j \overline{\xi_k} f(t_j - t_k) \geq 0.$$

Sufficiency We suppose that (4.2.5) holds if (4.2.6) is satisfied. Then

$$\sum_{j=1}^{m}\sum_{k=1}^{m}\xi_j\overline{\xi_k}f(t_j-t_k) = |\xi_1|^2 f(0) + \sum_{j=2}^{m}\sum_{k=2}^{m}\xi_j\overline{\xi_k}(f(t_j-t_k)$$
$$-f(t_j-t_1)-f(t_1-t_k)) \geq 0.$$

Letting $t_1 = 0$, since $f(0) = 0$, we obtain f as negative definite.

We give now a characterization of infinitely divisible characteristic functions which is very similar to the characterization of arbitrary characteristic functions given by the Bochner theorem.

Theorem 4.2.6 (*Schoenberg's theorem*) f is the logarithm of an infinitely divisible characteristic function if and only if f is a hermitian, continuous, negative-definite function satisfying $f(0) = 0$.

Proof *Necessity* If $\hat{p}$ is an infinitely divisible characteristic function and if $f = \log \hat{p}$, it is clear that f is continuous, hermitian, and satisfies $f(0) = 0$. Now, from Corollary 4.2.2 and Bochner's theorem, $e^{\alpha f}$ is a nonnegative-definite function for any $\alpha > 0$, so that

$$\sum_{j=1}^{m}\sum_{k=1}^{m}\xi_j\overline{\xi_k}e^{\alpha f(t_j-t_k)} \geq 0$$

for any $\alpha > 0$, any nonnegative integer m, any complex constants $\xi_1, \ldots, \xi_m$, and any $t_1, \ldots, t_m$ belonging to R^n. If (4.2.6) is satisfied, then we have for any $\alpha > 0$

$$\sum_{j=1}^{m}\sum_{k=1}^{m}\xi_j\overline{\xi_k}(e^{\alpha f(t_j-t_k)}-1)\alpha^{-1} \geq 0.$$

Letting $\alpha \to 0$, we obtain (4.2.5) and, from Lemma 4.2.1, f is negative definite.

Sufficiency We suppose that (4.2.4) is satisfied. If we consider the matrix A with elements $a_{j,k}$ defined by

$$a_{j,k} = f(t_j - t_k) - f(t_j) - f(-t_k),$$

then A is a nonnegative-definite hermitian matrix. It is well known [cf. for example, Marcus and Minc (1965, Section 3.4)] that there exists an orthonormal matrix with elements $c_{j,k}$ which reduces A to a diagonal matrix having nonnegative elements b_j in the diagonal. In other words

$$a_{j,k} = \sum_{l=1}^{m} c_{j,l}\overline{c_{k,l}}\, b_l$$

for any j and k, that is,

$$f(t_j - t_k) = f(t_j) + \overline{f(t_k)} + \sum_{l=1}^{m} c_{j,l}\overline{c_{k,l}}\, b_l\,.$$

Now

$$S = \sum_{j=1}^{m} \sum_{k=1}^{m} \xi_j \overline{\xi_k} e^{\alpha f(t_j - t_k)}$$

$$= \sum_{j=1}^{m} \sum_{k=1}^{m} \eta_j \overline{\eta_k} \exp\left(\sum_{l=1}^{m} c_{j,l} \overline{c_{k,l}} b_l \right)$$

if

$$\eta_j = \xi_j \exp(\alpha f(t_j))$$

($j = 1, \ldots, m$). From this we obtain

$$S = \sum_{k_1=0}^{\infty} \cdots \sum_{k_m=0}^{\infty} \left| \sum_{j=1}^{m} \eta_j c_{j,1}^{k_1} \cdots c_{j,m}^{k_m} \right|^2 \frac{(b_1)^{k_1} \cdots (b_m)^{k_m}}{k_1! \cdots k_m!} \geq 0.$$

Therefore $e^{\alpha f}$ is a nonnegative-definite function for any $\alpha > 0$ and, from Bochner's theorem, $\hat{p}^{\alpha}$ is a characteristic function for any $\alpha > 0$. Therefore f is the logarithm of an in infinitely divisible characteristic function.

4.3 CANONICAL REPRESENTATIONS

We introduce now the following definitions.†

If $\alpha \in R^n$, P is a nonnegative quadratic form on R^n, and μ is a finite signed measure having no atom at the origin, a characteristic function $\hat{p}$ without zeros admits a (α, P, μ)-Lévy–Hinčin representation if

$$\log \hat{p}(t) = i(\alpha, t) - P(t) + \int K(t, u)\mu(du) \qquad (t \in R^n)$$

where K is defined by

$$K(t, u) = \begin{cases} \left(e^{i(t,u)} - 1 - \dfrac{i(t, u)}{1 + \|u\|^2} \right) \dfrac{1 + \|u\|^2}{\|u\|^2} & \text{if} \quad u \neq 0, \\ 0 & \text{if} \quad u = 0. \end{cases}$$

μ is the Poisson measure of p and the support of μ is the Poisson spectrum of p. If μ is nonnegative, the normal probability $\mathfrak{n}(0, P)$ is the normal factor of p. (If $P = 0$, then p has no normal factor.)

If $\alpha \in R^n$, P is a nonnegative quadratic form on R^n and μ is a finite signed measure having no atom at the origin, a characteristic function $\hat{p}$ without zeros admits a (α, P, μ)-De Finetti representation if

$$\log \hat{p}(t) = i(\alpha, t) - P(t) + \int (e^{i(t,u)} - 1)\mu(du) \tag{4.3.1}$$

for any $t \in R^n$.

† It must be noted that this terminology is different from that given in the case $n = 1$ for infinitely divisible characteristic functions by Lukacs (1970).

For example, $\hat{\delta}_\alpha$ admits a $(\alpha, 0, 0)$-Lévy–Hinčin representation and a $(\alpha, 0, 0)$-De Finetti representation. The normal characteristic function $\hat{\mathfrak{n}}(\alpha, P)$ admits a $(\alpha, P, 0)$-Lévy–Hinčin representation and a $(\alpha, P, 0)$-De Finetti representation. The characteristic function of the Poisson-type probability $\mathfrak{p}(\alpha, m, l)$ admits a $(\alpha, 0, \mu)$-De Finetti representation with

$$\mu(B) = \sum_{(\varepsilon_1 m_1, \ldots, \varepsilon_n m_n) \in B} l_{\varepsilon_1, \ldots, \varepsilon_n}$$

for any $B \in \mathscr{B}_n$.

We first prove that the De Finetti representation is a particular case of the Lévy–Hinčin representation with

Theorem 4.3.1 Let $p \in \mathscr{P}_n$. If $\hat{p}$ admits a (α, P, μ)-De Finetti representation, then it admits a (β, P, ν)-Lévy–Hinčin representation where

$$\nu(B) = \int_B \frac{\|u\|^2}{1 + \|u\|^2} \mu(du) \tag{4.3.2}$$

for any $B \in \mathscr{B}_n$ and

$$(\alpha, t) = (\beta, t) + \int \frac{(t, u)}{1 + \|u\|^2} \mu(du) \tag{4.3.3}$$

for any $t \in R^n$. Conversely, if $\hat{p}$ admits a (β, P, ν)-Lévy–Hinčin representation with a Poisson measure ν satisfying

$$\int \frac{1 + \|u\|^2}{\|u\|^2} \nu(du) < +\infty, \tag{4.3.4}$$

then $\hat{p}$ admits a (α, P, μ)-De Finetti representation where μ and α are defined by (4.3.2) and (4.3.3).

Proof If ν and β are defined by (4.3.2) and (4.3.3), then μ and α are given by

$$\mu(B) = \int_B \frac{1 + \|u\|^2}{\|u\|^2} \nu(du)$$

for any $B \in \mathscr{B}_n$ and

$$(\alpha, t) = (\beta, t) - \int \frac{(t, u)}{\|u\|^2} \nu(du)$$

for any $t \in R^n$ and this implies easily the theorem.

It must be noted that the measure ν defined by (4.3.2) has the same support as μ, so that if a characteristic function $\hat{p}$ admits a (α, P, μ)-De Finetti representation, the support of μ is the Poisson spectrum of p. If 0 does not belong to the Poisson spectrum of p, then the condition (4.3.4) is always satisfied, so that we have

Corollary 4.3.1 If $\hat{p}$ admits a (α, P, μ)-Lévy–Hinčin representation and if 0 does not belong to the support of μ, then $\hat{p}$ admits a De Finetti representation.

Theorem 4.3.2 Let E be some subspace of R^n and $p \in \mathscr{P}_n$. If $\hat{p}$ admits a (α, P, μ)-Lévy–Hinčin representation, then $\widehat{\mathrm{pr}_E\, p}$ admits a (β, Q, ν)-Lévy–Hinčin representation where β, Q, and ν are defined by

$$(\beta, t) = (\mathrm{pr}_E\, \alpha, t) + \int (t, \mathrm{pr}_E\, u) \frac{\|u\|^2 - \|\mathrm{pr}_E\, u\|^2}{1 + \|\mathrm{pr}_E\, u\|^2} \frac{\mu(du)}{\|u\|^2}$$

for any $t \in R^n$,

$$Q = \mathrm{pr}_E\, P, \qquad \nu = \mathrm{pr}_E\, \mu',$$

with

$$\mu'(B) = \int_B \frac{\|\mathrm{pr}_E\, u\|^2}{1 + \|\mathrm{pr}_E\, u\|^2} \frac{1 + \|u\|^2}{\|u\|^2} \mu(du) \tag{4.3.5}$$

for any $B \in \mathscr{B}_n$. If $\hat{p}$ admits a (α, P, μ)-De Finetti representation, then $\widehat{\mathrm{pr}_E\, p}$ admits a (β, Q, ν)-De Finetti representation where

$$\beta = \mathrm{pr}_E\, \alpha, \qquad Q = \mathrm{pr}_E\, P, \qquad \nu = \mathrm{pr}_E\, \mu - \mu(E)\delta_0\, .$$

Proof If $\hat{p}$ admits a (α, P, μ)-Lévy–Hinčin representation, then

$$\begin{aligned}
\hat{p}(\mathrm{pr}_E\, t) &= i(\alpha, \mathrm{pr}_E\, t) - P(\mathrm{pr}_E\, t) \\
&\quad + \int \left(e^{i(\mathrm{pr}_E t, u)} - 1 - \frac{i(\mathrm{pr}_E\, t, u)}{1 + \|u\|^2}\right) \frac{1 + \|u\|^2}{\|u\|^2} \mu(du) \\
&= i(\mathrm{pr}_E\, \alpha, t) - \mathrm{pr}_E\, P(t) \\
&\quad + \int \left(e^{i(t, \mathrm{pr}_E u)} - 1 - \frac{i(t, \mathrm{pr}_E\, u)}{1 + \|u\|^2}\right) \frac{1 + \|u\|^2}{\|u\|^2} \mu(du) \\
&= i(\mathrm{pr}_E\, \alpha, t) + i \int \frac{(t, \mathrm{pr}_E\, u)}{1 + \|\mathrm{pr}_E\, u\|^2} \\
&\quad \times (\|u\|^2 - \|\mathrm{pr}_E\, u\|^2) \frac{\mu(du)}{\|u\|^2} \\
&\quad - \mathrm{pr}_E\, P(t) + \int \left(e^{i(t, \mathrm{pr}_E u)} - 1 - \frac{i(t, \mathrm{pr}_E\, u)}{1 + \|\mathrm{pr}_E\, u\|^2}\right) \\
&\quad \times \frac{1 + \|\mathrm{pr}_E\, u\|^2}{\|\mathrm{pr}_E\, u\|^2} \mu'(du)
\end{aligned}$$

where μ' is defined by (4.3.5). (It is clear that μ' is a finite signed measure having no atom at the origin.) The first assertion of the theorem follows from this formula and from projection theorem. The second assertion can be proved exactly as the projection theorem.

We prove now the uniqueness of these representations.

Theorem 4.3.3 If an n-variate characteristic function $\hat{p}$ admits an (α, P, μ)-Lévy–Hinčin representation, then α, P, and μ are uniquely determined by p.

Proof We consider first the case $n = 1$. If we have

$$\log \hat{p}(t) = \phi(t) = i\alpha t - \gamma t^2 + \int K(t, u)\mu(du),$$

then

$$-\tfrac{1}{2}(\phi(t) + \phi(-t)) = \gamma t^2 + \int (1 - \cos(tu))\frac{1 + u^2}{u^2}\mu(du). \tag{4.3.6}$$

If $|u| < 1/t^\rho$ and if $\rho > 1$, then

$$\lim_{t \to \infty} \frac{(1 - \cos(tu))(1 + u^2)}{t^2 u^2} = \frac{1}{2}$$

implies

$$\lim_{t \to \infty} \frac{1}{t^2} \int_{\{|u| < 1/t^\rho\}} (1 - \cos(tu))\frac{1 + u^2}{u^2}\mu(du) = 0. \tag{4.3.7}$$

If $|u| \geq 1/t^\sigma$, then

$$(1 - \cos(tu))\frac{1 + u^2}{u^2} < 2(1 + t^{2\sigma}),$$

so that

$$\lim_{t \to \infty} \frac{1}{t^2} \int_{\{|u| \geq 1/t^\sigma\}} (1 - \cos(tu))\frac{1 + u^2}{u^2}\mu(du) = 0 \tag{4.3.8}$$

if $\sigma < 1$. Now if $1/t^\rho \leq |u| < 1/t^\sigma$, then

$$(1 - \cos(tu))\frac{1 + u^2}{u^2} < 2(1 + t^{2\rho})$$

and since

$$|\mu|(\{1/t^\rho \leq |u| < 1/t^\sigma\}) = O(1/t^\sigma),$$

then

$$\lim_{t \to \infty} \frac{1}{t^2} \int_{\{1/t^\rho \leq |u| < 1/t^\sigma\}} (1 - \cos(tu))\frac{1 + u^2}{u^2}\mu(du) = 0 \tag{4.3.9}$$

if $2(\rho - 1) < \sigma$. From (4.3.6)–(4.3.9), we obtain easily

$$\gamma = -\frac{1}{2} \lim_{t \to \infty} \frac{\phi(t) + \phi(-t)}{t^2},$$

and γ is determined uniquely by ϕ.

We go now to the general case. If

$$\log \hat{p}(t) = \phi(t) = i(\alpha, t) - P(t) + \int K(t, u)\mu(du),$$

we obtain from Theorem 4.3.2 and the case $n = 1$ that P is determined uniquely by ϕ. Now, if

$$\begin{aligned} \psi(t, h) &= \phi(t) - \tfrac{1}{2}(\phi(t + h) + \phi(t - h)) - P(h) \\ &= \int e^{i(t, u)}(1 - \cos(h, u)) \frac{1 + \|u\|^2}{\|u\|^2} \mu(du) \end{aligned}$$

and

$$l(t) = \int_C \psi(t, h)\lambda(dh)$$

where $C = \{x = (x_1, \ldots, x_n) \in R^n : 0 \leq x_j \leq 1, j = 1, \ldots, n\}$, then

$$l(t) = \int e^{i(t, u)}\left(1 - \int_C \cos(h, u)\lambda(dh)\right)\left(\frac{1 + \|u\|^2}{\|u\|^2}\right) \mu(du),$$

and $l = \hat{v}$ where v is the measure defined by

$$v(B) = \int_B \left(1 - \int_C \cos(h, u)\lambda(dh)\right) \frac{1 + \|u\|^2}{\|u\|^2} \mu(du)$$

for any $B \in \mathscr{B}_n$. From this, we have

$$\mu(B) = \int_B \left(1 - \int_D \cos(h, u)\lambda(dh)\right)^{-1} \frac{\|u\|^2}{1 + \|u\|^2} v(du)$$

for any $B \in \mathscr{B}_n$. Since l is determined uniquely by ϕ and since $l = \hat{v}$, then v is determined uniquely by ϕ, so that μ is determined uniquely by ϕ. Finally, α is determined uniquely by ϕ and the theorem is proved.

From Theorems 4.3.1 and 4.3.3, we obtain easily

Corollary 4.3.2 If an n-variate characteristic function $\hat{p}$ admits a (α, P, μ)-De Finetti representation, then α, P, and μ are uniquely determined by p.

From Theorems 4.3.2 and 4.3.3, we have also

Corollary 4.3.3 Let E be some subspace of R^n and $p \in \mathscr{P}_n$. If $\hat{p}$ admits a (α, P, μ)-Lévy–Hinčin representation, then $\mathrm{pr}_E\, p$ is a normal probability if and only if μ is concentrated on $E^\perp$.

Theorem 4.3.4 If $\mu \in \mathscr{M}_n$, there exists a characteristic function $\hat{p}$ admitting a $(\alpha, 0, \mu)$-De Finetti representation if and only if $\exp \mu$ is a measure. Then p is given by

$$p = \frac{\delta_\alpha * \exp \mu}{\exp(\mu(R^n))}. \tag{4.3.10}$$

Proof If the characteristic function $\hat{p}$ admits a representation (4.3.1) with $P = 0$, then

$$\begin{aligned}\hat{p}(t) &= e^{i(\alpha, t)} \exp\left[\int (e^{i(t, u)} - 1)\mu(du)\right] \\ &= \frac{e^{i(\alpha, t)}}{\exp(\mu(R^n))}\left[\sum_{k=0}^{\infty} \frac{1}{k!} e^{i(t, u)}\mu(du)\right]^k.\end{aligned}$$

From this relation and from the convolution and continuity theorems, we have

$$\begin{aligned}\hat{p}(t) &= \int e^{i(t, u)} p(du) \\ &= \frac{e^{i(\alpha, t)}}{\exp(\mu(R^n))}\int e^{i(t, u)}\left(\sum_{k=0}^{\infty} \frac{1}{k!}\mu^{*k}(du)\right) \\ &= \int e^{i(t, u)} \frac{\delta_\alpha * \exp \mu}{\exp(\mu(R^n))}(du)\end{aligned}$$

which is, by the uniqueness theorem, equivalent to (4.3.10) and proves the necessity of the condition. Conversely, if $\exp \mu$ is a measure, then the same calculation proves that the characteristic function $\hat{p}$ defined by (4.3.10) admits representation (4.3.1) with $P = 0$.

Corollary 4.3.4 If $\mu \in \mathscr{M}_n$ is concentrated on $A \in F_\sigma^n$ and if the characteristic function $\hat{p}$ admits a $(\alpha, 0, \mu)$-De Finetti representation, then p is concentrated on $(N)A + \alpha$.

Corollary 4.3.5 If $\alpha \in R^n$, P is a nonnegative quadratic form, and μ is a finite measure, there exists a characteristic function admitting a (α, P, μ)-De Finetti representation and this characteristic function is infinitely divisible.

Proof If μ is a measure, $\exp \mu$ is a measure and, from Theorem 4.3.4, the convolution of the probability p defined by (4.3.10) and the normal probability $\mathfrak{n}(0, P)$ have the required property. Since it is true also for α/k, P/k, and μ/k for any positive integer k, this convolution is infinitely divisible.

Corollary 4.3.6 If $\alpha \in R^n$, P is a nonnegative quadratic form and μ is a finite measure, there exists a characteristic function admitting a (α, P, μ)-Lévy–Hinčin representation and this characteristic function is infinitely divisible.

Proof Let ν_j be the measure defined by

$$\nu_j(B) = \int_B \frac{1 + \|u\|^2}{\|u\|^2} \mu_j(du)$$

for any $B \in \mathscr{B}_n$ where

$$\mu_j(B) = \mu(B \cap \{x \in R^n : \|x\| > 1/j\})$$

($j = 1, 2, \ldots$). Then, from Corollary 4.3.5, there exists an infinitely divisible characteristic function $\hat{p}_j$ admitting a (α, P, ν_j)-De Finetti representation and, from Theorem 4.3.1, there exists an infinitely divisible characteristic function $\hat{q}_j$ admitting a (α, P, μ_j)-Lévy–Hinčin representation. When $j \to \infty$, $\hat{q}_j$ tends to the function f defined by

$$\log f(t) = i(\alpha, t) - P(t) + \int K(t, u)\mu(du).$$

Since f is continuous, it follows from the continuity theorem and Theorem 4.2.3 that f is the characteristic function of an infinitely divisible probability.

Corollary 4.3.7 Let $\hat{p}$ and $\hat{q}$ be two characteristic functions admitting a (α, P, μ)-Lévy–Hinčin representation (resp. De Finetti representation) and a (β, Q, ν)-Lévy–Hinčin representation (resp. De Finetti representation). If $Q \leq P$ and $\nu \leq \mu$, then q divides p.

Proof From Corollary 4.3.6 (resp. 4.3.5), there exists an infinitely divisible characteristic function $\hat{r}$ admitting a $(\alpha - \beta, P - Q, \mu - \nu)$-Lévy–Hinčin (resp. De Finetti) representation. Since $\hat{p}\hat{q} = \hat{r}$, q divides p.

Theorem 4.3.5 Let $\{\hat{p}_j\}$ be a sequence of characteristic functions admitting a (α_j, P_j, μ_j)-Lévy–Hinčin representation with some finite measures μ_j having no atom at the origin ($j = 1, 2, \ldots$). Then $\{p_j\}$ converges completely to p if and only if $\hat{p}$ admits a (α, P, μ)-Lévy–Hinčin representation with

(a) $\lim_{j\to\infty} \alpha_j = \alpha$,

(b) $$\lim_{m\to 0} \limsup_{j\to\infty} \left[P_j(t) + \int_{\{\|u\| \leq m\}} \frac{(t, u)^2}{2} \frac{1 + \|u\|^2}{\|u\|^2} \mu_j(du) \right]$$
$$= \lim_{m\to 0} \liminf_{j\to\infty} \left[P_j(t) + \int_{\{\|u\| \leq m\}} \frac{(t, u)^2}{2} \frac{1 + \|u\|^2}{\|u\|^2} \mu_j(du) \right]$$
$$= P(t),$$

(c) $\lim_{j\to\infty} \mu_j^m \overset{c}{=} \mu^m$,

where μ_j^m and μ^m are the restrictions† of μ_j and μ to $D_m = \{x \in R^n : \|x\| > m\}$ for any $m > 0$ such that D_m is a continuity set of μ.

Proof *Necessity* From Corollary 4.3.6, p_j is infinitely divisible and it follows from Theorem 4.2.3 that p is infinitely divisible so that $\log \hat{p} = \phi$ exists and

$$\log \hat{p}_j(t) = \phi_j(t) \to \phi(t)$$

when $j \to \infty$, the convergence being uniform for t belonging to a compact set of R^n.

If $C_h = \{u = (u_1, \ldots, u_n) \in R^n : -h < u_k < h,\ k = 1, \ldots, n\}$, we obtain

$$\begin{aligned} -\int_{C_h} &\phi_j(t)\lambda(dt) \\ &= (2^n/3)h^{n+2} \sum_{l=1}^{n} \gamma_{j,l} + (2h)^n \int \left(1 - \prod_{l=1}^{n} \frac{\sin(hu_l)}{hu_l}\right) \frac{1 + \|u\|^2}{\|u\|^2} \mu_j(du) \end{aligned} \tag{4.3.11}$$

where $\gamma_{j,l}$ is the coefficient of t_l^2 in $P_j(t)$. Since the left side of (4.3.11) tends to $-\int_{C_h} \phi(t)\lambda(dt)$ when $j \to \infty$ and since the two terms of the right side are positive, these two terms are bounded. In particular

$$\int \left(1 - \prod_{l=1}^{n} \frac{\sin(hu_l)}{hu_l}\right) \frac{1 + \|u\|^2}{\|u\|^2} \mu_j(du)$$

is bounded. Since there exist some constants K_1 and K_2 such that

$$0 < K_1 \leq \left(1 - \prod_{l=1}^{n} \frac{\sin(hu_l)}{hu_l}\right) \frac{1 + \|u\|^2}{\|u\|^2} < K_2 < +\infty,$$

we find that $\{\mu_j\}$ is uniformly bounded.

From (4.3.11), we obtain

$$\begin{aligned} \int &\left(1 - \prod_{l=1}^{n} \frac{\sin(hu_l)}{hu_l}\right) \frac{1 + \|u\|^2}{\|u\|^2} \mu_j(du) \\ &= -(1/(2h)^n) \int_{C_h} \phi_j(t)\lambda(dt) - \tfrac{1}{3}h^2 \sum_{l=1}^{n} \gamma_{j,l}. \end{aligned}$$

---

† The restriction v of a signed measure μ to a Borel set A is defined by $v(B) = \mu(B \cap A)$ for any $B \in \mathscr{B}_n$.

Since ϕ_j converges to ϕ uniformly on C_h and since $\phi(0) = 0$, we obtain the existence of some j_0 and h_0 such that

$$\int \left(1 - \prod_{l=1}^{n} \frac{\sin(h_0 u_l)}{h_0 u_l}\right) \frac{1 + \|u\|^2}{\|u\|^2} \mu_j(du) < \frac{1}{2} \varepsilon$$

if $j > j_0$. Since for $\|u\| > 2\sqrt{n}\, h_0$

$$\left(1 - \prod_{l=1}^{n} \frac{\sin(h_0 u_l)}{h_0 u_l}\right) \frac{1 + \|u\|^2}{\|u\|^2} > \frac{1}{2},$$

we obtain

$$\mu_j(\{u : \|u\| > 2\sqrt{n}\, h_0\}) < \varepsilon$$

for $j > j_0$ and therefore $\{\mu_j\}$ is tight. From Theorem 2.6.1, we deduce the existence of a subsequence $\{\mu_{j_k}\}$ of $\{\mu_j\}$ which converges completely to some measure μ. It follows from Theorem 2.6.3 that

$$\lim_{k\to\infty} \mu_{j_k}^m \overset{c}{=} \mu^m \tag{4.3.12}$$

where $\mu_{j_k}^m$ and μ^m are the restrictions of μ_{j_k} and μ to D_m for any m such that D_m is a continuity set of μ.

If we define $I_{j,m}$ by

$$I_{j,m}(t) = \int_{\{\|u\| \le m\}} \left(e^{i(t,u)} - 1 - \frac{i(t,u)}{1 + \|u\|^2} + \frac{(t,u)^2}{2}\right) \frac{1 + \|u\|^2}{\|u\|^2} \mu_j(du),$$

we obtain

$$\varphi_j(t) = i(\alpha_j, t) - P_j(t) + I_{j,m}(t) + \int_{\{\|u\| > m\}} K(t,u)\mu_j(du)$$
$$- \int_{\{\|u\| > m\}} \frac{(t,u)^2}{2} \frac{1 + \|u\|^2}{\|u\|^2} \mu_j(du). \tag{4.3.13}$$

Since

$$\lim_{m\to 0} \sup_{\|u\| \le m} \left| e^{i(t,u)} - 1 - \frac{i(t,u)}{1 + \|u\|^2} + \frac{(t,u)^2}{2} \right| \frac{1 + \|u\|^2}{\|u\|^2} = 0,$$

we obtain

$$\begin{aligned} \lim_{m\to 0} \limsup_{j\to\infty} \operatorname{Re} I_{j,m}(t) &= \lim_{m\to 0} \liminf_{j\to\infty} \operatorname{Re} I_{j,m}(t) = 0, \\ \lim_{m\to 0} \limsup_{j\to\infty} \operatorname{Im} I_{j,m}(t) &= \lim_{m\to 0} \liminf_{j\to\infty} \operatorname{Im} I_{j,m}(t) = 0. \end{aligned} \tag{4.3.14}$$

From the dominated convergence theorem, we have

$$\lim_{k\to\infty} \int_{\{\|u\|>m\}} K(t, u)\mu_{j_k}(du) = \int_{\{\|u\|>m\}} K(t, u)\mu(du). \tag{4.3.15}$$

Taking the real part of the two sides of (4.3.13), we obtain from (4.3.14), (4.3.15), and the convergence of $\{\phi_j\}$ to ϕ,

$$\operatorname{Re}\phi(t) = \int \operatorname{Re} K(t, u)\mu(du) - \lim_{m\to 0} \limsup_{k\to\infty} \times \left(P_{j_k}(t) + \int_{\{\|u\|\le m\}} \frac{(t, u)^2}{2} \frac{1 + \|u\|^2}{\|u\|^2} \mu_{j_k}(du)\right) \tag{4.3.16}$$

and the same relation with lim inf instead of lim sup, so that

$$\begin{aligned} &\lim_{m\to 0} \limsup_{k\to\infty} \left(P_{j_k}(t) + \int_{\{\|u\|\le m\}} \frac{(t, u)^2}{2} \frac{1 + \|u\|^2}{\|u\|^2} \mu_{j_k}(du)\right) \\ &= \lim_{m\to 0} \liminf_{k\to\infty} \left(P_{j_k}(t) + \int_{\{\|u\|\le m\}} \frac{(t, u)^2}{2} \frac{1 + \|u\|^2}{\|u\|^2} \mu_{j_k}(du)\right) = P(t) \end{aligned} \tag{4.3.17}$$

for a nonnegative quadratic form P. In the same way, we obtain

$$\operatorname{Im}\phi(t) = \int \operatorname{Im} K(t, u)\mu(du) + \lim_{k\to\infty} (\alpha_{j_k}, t), \tag{4.3.18}$$

so that

$$\lim_{k\to\infty} \alpha_{j_k} = \alpha \tag{4.3.19}$$

for some $\alpha \in R^n$.

From (4.3.16) and (4.3.18), we deduce that $\hat{p}$ admits an (α, P, μ)-Lévy–Hinčin representation and from Theorem 4.3.3 it follows that α, P, and μ are determined by ϕ. Therefore the limits in (4.3.12), (4.3.17), and (4.3.19) are independent of the subsequence $\{p_{j_k}\}$ and this implies easily the necessity of the condition of the theorem.

Sufficiency From Corollary 4.3.6, there exists a characteristic function $\hat{p}$ admitting an (α, P, μ)-Lévy–Hinčin representation. From conditions (a)–(c), we deduce easily that $\{\hat{p}_j\}$ converges to $\hat{p}$ and, by the continuity theorem, $\{p_j\}$ converges completely to p.

We can now prove an important result for infinitely divisible probabilities.

Theorem 4.3.6 (*Lévy–Hinčin representation theorem*) An n-variate probability p is infinitely divisible if and only if $\hat{p}$ admits a (α, P, μ)-Lévy–Hinčin representation with some finite measure μ.

Proof The sufficiency follows from Corollary 4.3.6. Conversely, if p is an infinitely divisible probability, from De Finetti's theorem, there exists a sequence of probabilities $\{p_j\}$ such that $\hat{p}_j$ admits a $(0, 0, \mu_j)$-De Finetti representation with some measures μ_j and $p_j \xrightarrow{c} p$. It follows then from Theorems 4.3.1 and 4.3.5 that $\hat{p}$ admits an (α, P, μ)-Lévy–Hinčin representation with some measure μ.

Another proof of this result can be deduced from Schoenberg's theorem and from Choquet's theory on the representation of cones. For such a proof, see Kendall (1963) and Johansen (1966).

From Theorems 4.3.6, 4.3.1, and 4.3.3, we have

Corollary 4.3.8 If a characteristic function $\hat{p}$ admits an (α, P, μ)-Lévy–Hinčin representation (or De Finetti representation) with a signed measure μ which is negative on some subset of R^n, then p is not infinitely divisible.

Example Let a, b, and c be some elements of R^n satisfying $0 < 2a < b < c < 2b$. We denote by μ_j $(j = 1, 2)$ the measures defined by

$$\mu_j(B) = \lambda(B \cap A_j)$$

for any $B \in \mathscr{B}_n$ where $A_1 = \{x \in R^n : a \leq x \leq 2b\}$ and $A_2 = \{x \in R^n : b \leq x \leq c\}$ and by μ the signed measure

$$\mu = \mu_1 - (1 + \eta)\mu_2 .$$

We can choose $\eta > 0$ small enough so that μ^{*2}, μ^{*3}, $\mu + \frac{1}{2}\mu^{*2}$ and the projections $\mathrm{pr}_E\, \mu$ of μ on any subspace E of R^n $(E \neq R^n)$ are measures. Since for any integer $k > 1$, there exist some nonnegative integers l and m such that

$$k = 2l + 3m,$$

μ^{*k} is a measure for any $k > 1$. From

$$\exp \mu = \delta_0 + (\mu + \tfrac{1}{2}\mu^{*2}) + \sum_{k=3}^{\infty} (1/k!)\mu^{*k},$$

we deduce easily that $\exp \mu$ is a measure. If we consider the characteristic function $\hat{p}$ with $(\alpha, 0, \mu)$-De Finetti representation (its existence follows from Theorem 4.3.4), then p is not infinitely divisible (since μ is negative on A_2), but we deduce from Theorem 4.3.2 that all its projections are infinitely divisible.

Example Another similar example is the Wishart's probability p defined by

$$\hat{p}(t_1, t_2, t_3) = [(1 - it_1)(1 - it_3) + t_2^2]^{-1/2}.$$

Lévy (1948) has proved that p is indecomposable, but $\mathrm{pr}_\theta\, p$ is infinitely divisible for any $\theta \in R^n \backslash \{0\}$. Also the projections of p on the three coordinate planes are infinitely divisible. For the proof, we refer to the note by Lévy.

The study of characteristic functions admitting an (α, P, μ)-Lévy–Hinčin representation with a signed measure μ which is negative on some subset of R^n is more difficult. We give only some partial results in this case.

Theorem 4.3.7 If p is an n-variate probability satisfying

$$p(\{\alpha\}) > \tfrac{1}{2} \tag{4.3.20}$$

for some $\alpha \in R^n$, then $\hat{p}$ admits an $(\alpha, 0, \mu)$-De Finetti representation with a finite signed measure μ satisfying

$$\mu(R^n) < \log 2. \tag{4.3.21}$$

Moreover, if p is concentrated on $A + \alpha$ $(A \in F_\sigma^n)$, then μ is concentrated on $(N)A$ and

$$\mu(B) \geq 0$$

for any Borel set B satisfying

$$B \cap \bigcup_{k=2}^{\infty} (k)A = \varnothing. \tag{4.3.22}$$

Proof We prove the case $\alpha = 0$, the general case following easily from this particular one. If p satisfies

$$\omega = p(\{0\}) > \tfrac{1}{2},$$

and if we define q by

$$q(B) = \omega^{-1} p(B \backslash \{0\})$$

for any $B \in \mathscr{B}_n$, then q is a measure concentrated on A and satisfying $q(R^n) < 1$. From

$$\hat{p}(t) = \omega\left(1 + \int e^{i(t,\, u)} q(du)\right),$$

we deduce easily that

$$\begin{aligned} \log \hat{p}(t) &= \log \omega + \log\left(1 + \int e^{i(t,\, u)} q(du)\right) \\ &= \log \omega + \sum_{k=1}^{\infty} ((-1)^{k-1}/k)\left(\int e^{i(t,\, u)} q(du)\right)^k \\ &= \log \omega + \int e^{i(t,\, u)} \mu(du) \end{aligned}$$

where

$$\mu = \sum_{k=1}^{\infty} ((-1)^{k-1}/k) q^{*k}. \tag{4.3.23}$$

Since $\log \hat{p}(0) = 0$, we find

$$\log \omega = -\mu(R^n). \tag{4.3.24}$$

Therefore $\hat{p}$ admits a $(0, 0, \mu)$-De Finetti representation with μ defined by (4.3.23), and (4.3.21) follows easily from (4.3.24) and (4.3.20). From (4.3.23) we deduce that μ is concentrated on $(N)A$ and since

$$\mu - q = \sum_{k=2}^{\infty} ((-1)^{k-1}/k) q^{*k}$$

is concentrated on

$$\bigcup_{k=2}^{\infty} (k)A$$

we find that if B satisfies (4.3.22), then

$$\mu(B) = q(B) \geq 0.$$

Theorem 4.3.8 Let $k = (k_1, \ldots, k_n) \in R^n$. An n-variate characteristic function $\hat{p}$ admits an $(\alpha, 0, \mu)$-De Finetti representation with a lattice finite signed measure μ concentrated on $(Z)\{k_1 e_1, \ldots, k_n e_n\}$ and such that

$$D^{2(e_1 + \cdots + e_n)} \hat{\mu}(t)$$

exists for any $t \in R^n$ if and only if p is an n-variate probability satisfying the following conditions:

(a) p is a lattice probability concentrated on $(Z)\{k_1 e_1, \ldots, k_n e_n\} + \alpha$;
(b) $\inf_{t \in R^n} |\hat{p}(t)| > 0$;
(c) $D^{2(e_1 + \cdots + e_n)} \hat{p}(0)$ exists.

Proof If p is a lattice probability concentrated on $(Z)\{k_1 e_1, \ldots, k_n e_n\} + \alpha$, then $\hat{p}_\alpha = \hat{p} e^{-i(\alpha, \cdot)}$ is a function which is periodic with respect to the jth variable, the period being $2\pi/k_j$ if $k_j \neq 0$ and arbitrary if $k_j = 0$ ($j = 1, \ldots, n$). If $\hat{p}$ satisfies conditions (b) and (c), then $\phi = \log \hat{p}$ is a function which is periodic with the same period with respect to each variable and

$$D^{2(e_1 + \cdots + e_n)} \phi$$

exists and is continuous. It follows then from a well-known result of the theory of Fourier series† that ϕ is equal to its Fourier series and this is equivalent to the sufficiency of the condition of the theorem. The necessity follows immediately from Corollary 4.3.4.

---

† For a proof of this result in the case $n = 1$, see Bary (1964, p. 83). The general case can be proved by the same way.

In the case $n = 1$, we have also the following results.

Theorem 4.3.9 A univariate characteristic function $\hat{p}$ admits a $(0, 0, \mu)$-De Finetti representation with a purely atomic symmetric signed measure μ if and only if the following conditions are satisfied:

(a) p is a purely atomic symmetric probability;
(b) $\inf_{t \in R} \hat{p}(t) > 0$.

Moreover, if p is concentrated on A, then μ is concentrated on $(Z)A$.

Proof *Necessity* Condition (a) follows from (4.3.10) and (b) follows from

$$\hat{p}(t) \geq \exp(-2\|\mu\|).$$

Sufficiency If (a) is satisfied, $\hat{p}$ is an almost periodic function† with a summable Fourier series and we deduce from (b) and Theorem A.2.5 that $\log \hat{p}$ is also an almost periodic function with a summable Fourier series, that is, $\hat{p}$ admits a $(0, 0, \mu)$-De Finetti representation with a purely atomic signed measure μ. From Corollary 2.2.1, we deduce that $\log \hat{p}(t)$ is real for any $t \in R$, so that μ is symmetric. The last assertion follows from Theorem A.2.5.

Theorem 4.3.10 A univariate characteristic function $\hat{p}$ admits an $(\alpha, 0, \mu)$-De Finetti representation with a purely atomic finite signed measure μ satisfying

$$\int e^{-ru} |\mu| (du) < +\infty \tag{4.3.25}$$

for $|r| < \rho$ $(\rho > 0)$ if and only if the following conditions are satisfied:

(a) p is purely atomic;
(b) $\hat{p}$ is an analytic characteristic function which is regular and without zeros in the strip $S = \{t \in C : |\text{Im } t| < \rho\}$. Moreover, if p is concentrated on A, μ is concentrated on $(Z)A$.

Proof *Necessity* If $\hat{p}$ admits an $(\alpha, 0, \mu)$-De Finetti representation, then

$$p = (\delta_\alpha * \exp \mu)/\exp(\mu(R))$$

and p is purely atomic. If (4.3.25) is satisfied, then $\int e^{-ru}\mu^+(du)$ and $\int e^{-ru}\mu^-(du)$ converges for $|r| < \rho$. Since μ^+ and μ^- are probabilities up to a multiplicative constant factor, it follows from Theorem 3.3.2 that $\hat{\mu}^+$ and $\hat{\mu}^-$ are functions which are regular in S. Therefore, $\log \hat{p}$ is regular in S and this implies that $\hat{p}$ is regular and without zeros in S.

---

† For the properties of the almost periodic functions used in this section, see Appendix A.2.

Sufficiency Since $|\hat{p}|^2$ is the characteristic function of $p * \tilde{p}$ and is regular in S, the function ϕ_y defined by

$$\phi_y(t) = |\hat{p}(t + iy)|^2 \qquad (|y| < \rho)$$

is the characteristic function of a probability concentrated on $A + (-A) \subset (Z)A$ and is an almost periodic function with a summable Fourier series and satisfying

$$\Lambda_\phi \subset (Z)A.$$

Since $|\hat{p}|^2$ has no zeros in S, we deduce from Theorem A.2.6 that

$$\inf_{t \in R} |\phi_y(t)| > 0$$

and from Theorem A.2.5 that the function ψ_y defined by

$$\psi_y(t) = \log \phi_y(t)$$

is an almost periodic function with a summable Fourier series and satisfying

$$(Z)\Lambda_{\psi_y} \subset (Z)\Lambda_{\phi_y} \subset (Z)A.$$

From Corollary A.2.2, we deduce that

$$\psi_y(t) = \sum_{\lambda \in (Z)A} a_\lambda(y) e^{i\lambda t} \tag{4.3.26}$$

where $a_\lambda(y)$ are functions of y with complex values and satisfying

$$\sum_{\lambda \in (Z)A} a_\lambda(y) < +\infty$$

for any $|y| < \rho$.

If we consider the function u defined by

$$u(x, y) = \log |\hat{p}(x + iy)| = \tfrac{1}{2}\psi_y(x),$$

u is an almost periodic harmonic function in the strip $\{(x, y) \in R^2 : |y| < \rho\}$. Since $u(x, y) = u(-x, y)$, we deduce from Theorem A.2.7 that u admits a Fourier series

$$-\alpha y + \beta + \sum_{\lambda \in \Lambda} l_\lambda e^{-\lambda y} \cos(\lambda x) \tag{4.3.27}$$

where α, β, l_λ are real constants and Λ is a countable set. From the uniqueness of the Fourier series (Theorem A.2.3) of $\psi_y(x)$, we deduce from (4.3.26) and (4.3.27) that $\Lambda \subset (Z)A$ and

$$a_0(y) = -\alpha y + \beta, \qquad a_\lambda(y) = l_\lambda e^{-\lambda y} + l_{-\lambda} e^{\lambda y} \qquad \text{if } \lambda \in \Lambda.$$

This implies that $(l_\lambda e^{-\lambda y})_{\lambda \in \Lambda}$ is a summable family and from Theorem A.2.8

$$u(x, y) = -\alpha y + \beta + \sum_{\lambda \in \Lambda} l_\lambda e^{-\lambda y} \cos(\lambda x)$$

for $|y| < \rho$.

If we define the function g by

$$g(t) = i\alpha t + \beta + \sum_{\lambda \in (Z)A} l_\lambda e^{i\lambda t},$$

then g is a function which is regular in $\{t \in C: |\mathrm{Im}\, t| < \rho\}$. Since $\log \hat{p}$ is regular in the same strip and since

$$\mathrm{Re}\, g(t) = \log |\hat{p}(t)| = \mathrm{Re} \log \hat{p}(t),$$

then

$$\log \hat{p}(t) = g(t) + i\gamma$$

for some real constant γ. From $\log \hat{p}(0) = 0$, we deduce that $\gamma = 0$ and

$$\beta = -\sum_{l \in (Z)A} a_l\,,$$

so that

$$\log \hat{p}(t) = i\alpha t + \sum_{l \in (Z)A} a_l(e^{ilt} - 1)$$

and the theorem is proved.

We will prove in Section 7.3 the extension of these two results to the case of n-variate characteristic functions.

4.4 A LIMIT THEOREM

If $\{r_j\}$ is a sequence of positive integers, we define the *array* of probabilities as a set $\{p_{j,k} : k = 1, \ldots, r_j; j = 1, 2, \ldots\}$ where $p_{j,k} \in \mathscr{P}_n$ for each j and k.

We say that the array $\{p_{j,k}\}$ *converges* to p for the centering $\{a_j\}$ ($a_j \in R^n$, $j = 1, 2, \ldots$) if

$$\lim_{j \to \infty} (p_{j,1} * \cdots * p_{j,r_j} * \delta_{a_j}) = p.$$

Then p is the limit of the array $\{p_{j,k}\}$ for the centering $\{a_j\}$.

We say that the array $\{p_{j,k}\}$ is *null* if

$$\lim_{j \to \infty} \sup_{1 \le k \le r_j} p_{j,k}(\{x : \|x\| > \varepsilon\}) = 0$$

for any $\varepsilon > 0$.

Theorem 4.4.1 An array $\{p_{j,k}\}$ is null if and only if

$$\lim_{j\to\infty} \sup_{\substack{1\le k\le r_j\\ \|t\|\le m}} (1 - \hat{p}_{j,k}(t)) = 0 \tag{4.4.1}$$

for any $m > 0$.

Proof An array $\{p_{j,k}\}$ is null if and only if $\{p_{j,k}\}$ converges completely to δ_0 uniformly in k and, by the continuity theorem, this is equivalent to (4.4.1).

We introduce now a function τ defined on R^n with values in R^n:

$$\tau(u) = (\tau_1(u), \ldots, \tau_n(u)),$$

$$\tau_l(u) = \tau_l(u_1, \ldots, u_n) = \begin{cases} 1 & \text{if} \quad u_l > 1, \\ u_l & \text{if} \quad |u_l| \le 1 \quad (l = 1, \ldots, n), \\ -1 & \text{if} \quad u_l < -1. \end{cases}$$

Lemma 4.4.1 If

$$\int \tau_l(u)p(du) = 0 \qquad (l = 1, \ldots, n),$$

then

$$|1 - \hat{p}(t)| \le \left(2 + \sqrt{n}\,\|t\| + \frac{\|t\|^2}{2}\right) \int \|\tau(u)\|^2 p(du)$$

Proof

$$\hat{p}(t) - 1 = \int (e^{i(t,u)} - 1)p(du) = \int (e^{i(t,u)} - 1 - i(t, \tau(u)))p(du).$$

If $|u_l| \le 1$ $(l = 1, \ldots, n)$, we have

$$|e^{i(t,u)} - 1 - i(t, \tau(u))| \le \frac{1}{2}\left(\sum_{l=1}^{n} t_l u_l\right)^2 \le \tfrac{1}{2}\|t\|^2\|u\|^2 = \tfrac{1}{2}\|t\|^2\|\tau(u)\|^2.$$

If one of the $|u_l|$ is greater than 1, we have

$$|e^{i(t,u)} - 1 - i(t, \tau(u))| \le 2 + \sqrt{n}\,\|t\| \le (2 + \sqrt{n}\,\|t\|)\|\tau(u)\|^2$$

and the lemma follows easily from these estimations.

Lemma 4.4.2 For any probability p, there exists some $a = (a_1, \ldots, a_n) \in R^n$ such that $p_a = p * \delta_a$ satisfies

$$\int \tau_l(u)p_a(du) = 0 \qquad (l = 1, \ldots, n).$$

Proof This follows immediately from the fact that if $w = (w_1, \ldots, w_n) \in R^n$, then

$$\int \tau_l(u)(p * \delta_w)(du) = \int \tau_l(u - w)p(du) = \int \tau_l(u - w_l e_l)p(du)$$

and when w_l varies from $-\infty$ to $+\infty$, $\int \tau_l(u - w_l e_l)p(du)$ is a continuous function decreasing from $+1$ to -1.

We must note that in general a is not determined uniquely, but in the following we choose one of the a satisfying this condition.

We prove now

Theorem 4.4.2 A null array $\{p_{j,k}\}$ converges to p for the centering $\{a_j\}$ if and only if $\hat{p}$ admits an (α, P, μ)-Lévy–Hinčin representation with

$$\text{(a)}\quad \lim_{j\to\infty}\left((\alpha_j, t) + \sum_{k=1}^{r_j}\int \frac{(t, u)}{1 + \|u\|^2} q_{j,k}(du)\right) = (\alpha, t);$$

$$\text{(b)}\quad \lim_{m\to 0}\limsup_{j\to\infty}\left(\sum_{k=1}^{r_j}\int_{\{\|u\|\le m\}} \frac{(t, u)^2}{2} q_{j,k}(du)\right)$$

$$= \lim_{m\to 0}\liminf_{j\to\infty}\left(\sum_{k=1}^{r_j}\int_{\{\|u\|\le m\}} \frac{(t, u)^2}{2} q_{j,k}(du)\right)$$

$$= P(t);$$

$$\text{(c)}\quad \lim_{j\to\infty}\sum_{k=1}^{r_j}\int_{B\cap D_m} \frac{\|u\|^2}{1 + \|u\|^2} q_{j,k}(du) = \mu(B \cap D_m)$$

for any continuity set B of μ and any $m > 0$ such that $D_m = \{u : \|u\| > m\}$ is a continuity set of μ. Here

$$q_{j,k} = p_{j,k} * \delta_{b_{j,k}} \qquad (j = 1, 2, \ldots, \quad k = 1, \ldots, r_j)$$

is such that

$$\int \tau_l(u) q_{j,k}(du) = 0 \qquad (l = 1, \ldots, n)$$

and

$$\alpha_j = a_j + \sum_{k=1}^{r_j} b_{j,k} \qquad (j = 1, 2, \ldots).$$

Proof If $\{p_{j,k}\}$ is a null array, $\{q_{j,k}\}$ is also a null array and from Theorem 4.4.1, we have

$$\lim_{j\to\infty}\sup_{\substack{1\le k\le r_j\\ \|t\|\le m}} (1 - \hat{q}_{j,k}(t)) = 0 \tag{4.4.2}$$

for any $m > 0$. If $\{p_{j,k}\}$ converges to p for the centering $\{a_j\}$, then

$$\lim_{j \to \infty} e^{i(\alpha_j, t)} \prod_{k=1}^{r_j} \hat{q}_{j,k}(t) = \hat{p}(t). \tag{4.4.3}$$

Letting

$$\log \hat{p} = \phi, \qquad \log \hat{q}_{j,k} = \phi_{j,k}, \qquad \theta_{j,k} = \hat{q}_{j,k} - 1,$$

we obtain

$$\phi_{j,k}(t) = \theta_{j,k}(t)(1 + \rho_{j,k}(t)) \tag{4.4.4}$$

where

$$|\rho_{j,k}(t)| = \left| \sum_{l=1}^{\infty} \frac{(-1)^l (\theta_{j,k}(t))^l}{l+1} \right| \leq \frac{\theta_{j,k}(t)}{1 - \theta_{j,k}(t)}$$

and from (4.4.2)

$$\lim_{j \to \infty} \sup_{\substack{1 \leq k \leq r_j \\ \|t\| \leq m}} |\rho_{j,k}(t)| = 0$$

for any $m > 0$. Using Lemma 4.4.1, we find

$$\sup_{\|t\| \leq m} \sum_{k=1}^{r_j} |\theta_{j,k}(t)| = O(A_j) \qquad (j \to +\infty) \tag{4.4.5}$$

where

$$A_j = \sum_{k=1}^{r_j} \int \|\tau(u)\|^2 q_{j,k}(du).$$

From (4.4.4) and (4.4.5), we obtain

$$\sum_{k=1}^{r_j} \phi_{j,k}(t) = \sum_{k=1}^{r_j} \theta_{j,k}(t) + o(A_j) \qquad (j \to +\infty). \tag{4.4.6}$$

If $h > 0$ is such that $\hat{p}$ has no zeros in $C_h = \{u = (u_1, \ldots, u_n) \in R^n : -h \leq u_l \leq h, l = 1, \ldots, n\}$, from (4.4.3) and (4.4.6), we obtain

$$\sum_{k=1}^{r_j} \theta_{j,k}(t) + i(\alpha_j, t) + o(A_j) \to \phi(t)$$

for any $t \in C_h$ and by integrating this relation on C_h

$$\sum_{k=1}^{r_j} \int \left(1 - \prod_{m=1}^{n} \frac{\sin(hu_m)}{hu_m}\right) p_{j,k}(du) + o(A_j) \to -\frac{1}{(2h)^n} \int_{C_h} \phi(t) \lambda(dt). \tag{4.4.7}$$

If h is small enough,

$$1 - \prod_{m=1}^{n} \frac{\sin(hu_m)}{hu_m} \geq \frac{h^2\|u\|^2}{10} \geq \frac{h^2\|\tau(u)\|^2}{10},$$

so that

$$\sum_{k=1}^{r_j} \int \left(1 - \prod_{m=1}^{n} \frac{\sin(hu_m)}{hu_m}\right) p_{j,k}(du) \geq \frac{h^2}{10} A_j$$

and from (4.4.7) we deduce that

$$A_j = O(1) \qquad (j \to +\infty). \tag{4.4.8}$$

Since $\{q_{j,k}\}$ is a null array, $\log \hat{q}_{j,k}$ exists for any k if j is great enough and from (4.4.3), (4.4.6), and (4.4.8), we obtain

$$\lim_{j\to\infty} e^{i(\alpha_j, t)} \prod_{k=1}^{r_j} e^{(q_{j,k}(t) - 1)} = \hat{p}(t)$$

or

$$\lim_{j\to\infty} \left(\delta_{\alpha_j} * \frac{\exp q_{j,1}}{\exp\|q_{j,1}\|} * \cdots * \frac{\exp q_{j,r_j}}{\exp\|q_{j,r_j}\|}\right) \overset{c}{=} p$$

and, from Theorems 4.2.4, 4.2.2, and 4.2.3, we deduce that p is infinitely divisible and, from Theorem 4.3.6, $\hat{p}$ admits an (α, P, μ)-Lévy–Hinčin representation with some finite measure μ.

Noting that

$$e^{i(\alpha_j, \cdot)} \prod_{k=1}^{r_j} e^{(q_{j,k} - 1)}$$

admits a $(\beta_j, 0, \mu_j)$-Lévy–Hinčin representation with

$$(\beta_j, t) = (\alpha_j, t) + \sum_{k=1}^{r_j} \int \frac{(t, u)}{1 + \|u\|} q_{j,k}(du),$$

$$\mu_j(B) = \sum_{k=1}^{r_j} \int_B \frac{\|u\|^2}{1 + \|u\|^2} q_{j,k}(du) \qquad (B \in \mathscr{B}_n),$$

we deduce from Theorem 4.3.5 the other assertions of the theorem.

Corollary 4.4.1 p is an infinitely divisible probability if and only if p is the limit of a null array for some arbitrary centering.

Proof The sufficiency follows from the preceding theorem. We prove now the necessity of the condition. If p is an infinitely divisible probability, for any $j \in N$, there exists some probability p_j such that

$$\hat{p} = \hat{p}_j^j. \tag{4.4.9}$$

If we consider the triangular array $\{p_{j,k}\}$ defined by

$$p_{j,k} = p_j \qquad (k = 1, \ldots, j),$$

then p is the limit of the array $\{p_{j,k}\}$. Since $\hat{p}$ has no zeros, from (4.4.9), we deduce easily that

$$\lim_{j\to\infty} \sup_{\|t\| \le m} (1 - \hat{p}_j(t)) = 0$$

for any $m > 0$ and, from Theorem 4.4.1, $\{p_{j,k}\}$ is a null array.

4.5 HINČIN'S THEOREM

Let $m > 0$ and $\mathcal{Q}_m$ be the set of all the n-variate probabilities p such that $\hat{p}(t) \neq 0$ for $\|t\| \le m$. We note that if $m < m'$,

$$\mathcal{Q}_{m'} \subset \mathcal{Q}_m$$

and that

$$\mathcal{P}_n = \bigcup_m \mathcal{Q}_m .$$

Moreover, $\mathcal{Q}_m$ is factor-closed, that is, any factor of $p \in \mathcal{Q}_m$ belongs also to $\mathcal{Q}_m$.

We define on $\mathcal{Q}_m$ a functional H_m by

$$H_m(p) = -\int_{\{\|t\| \le m\}} \log |\hat{p}(t)| \lambda(dt).$$

Evident properties of H_m are

$$H_m(p) \ge 0, \tag{4.5.1}$$

$$H_m(p_1 * p_2) = H_m(p_1) + H_m(p_2), \tag{4.5.2}$$

$$\int_{\{\|t\| \le m\}} (1 - |\hat{p}(t)|)\lambda(dt) \le H_m(p), \tag{4.5.3}$$

$$H_m(p) = 0 \qquad \text{if and only if } p \text{ is degenerate.} \tag{4.5.4}$$

We will use an extension of this last property:

Theorem 4.5.1 If $\{p_j\}$ is a sequence of probabilities belonging to $\mathcal{Q}_m$ and if

$$\lim_{j\to\infty} H_m(p_j) = 0,$$

there exists a sequence $\{\alpha_j\}$ of elements of R^n such that

$$\lim_{j\to\infty} (p_j * \delta_{\alpha_j}) = \delta_0 .$$

Proof We have from (4.5.2) and (4.5.3)

$$\int_{\{\|t\| \le m\}} (1 - |\hat{p}_j(t)|^2)\lambda(dt) \le H_m(p_j * \tilde{p}_j) = 2H_m(p_j) \tag{4.5.5}$$

and from Theorem 2.1.3

$$\int_{\{\|t\| \le 2m\}} (1 - |\hat{p}_j(t)|^2)\lambda(dt) \le 2^{n+2} \int_{\{\|t\| \le m\}} (1 - |\hat{p}_j(t)|^2)\lambda(dt)$$

and by iteration

$$\int_{\{\|t\| \le 2^k m\}} (1 - |\hat{p}_j(t)|^2)\lambda(dt) \le 2^{k(n+2)} \int_{\{\|t\| \le m\}} (1 - |\hat{p}_j(t)|^2)\lambda(dt). \tag{4.5.6}$$

From (4.5.5), (4.5.6), and the hypotheses, we deduce that

$$\lim_{j\to\infty} |\hat{p}_j(t)|^2 = 1 \qquad (t \in R^n)$$

and, from the continuity theorem,

$$\lim_{j\to\infty} (p_j * \tilde{p}_j) \stackrel{c}{=} \delta_0 .$$

The theorem follows then from the following lemma.

Lemma 4.5.1 If q divides p, then for any $l > 0$

$$\sup_{\xi \in R^n} p(\{x : \|x - \xi\| < l\}) \le \sup_{\xi \in R^n} q(\{x : \|x - \xi\| < l\}).$$

Proof If $p = q * r$, $A_y = \{x : \|x - y\| < l\}$, and $Q = \sup_{\xi \in R^n} q(\{x : \|x - \xi\| < l\})$, we have

$$p(A_y) = \int q(A_y - z)r(dz) = \int q(A_{y-z})r(dz) \le Q \int r(dz) = Q$$

for any y and this implies the lemma.

We can now prove

Theorem 4.5.2 (*Hinčin's theorem*) An n-variate probability p is always the convolution of two (perhaps degenerate) probabilities p_1 and p_2, where p_1 is a probability which has no indecomposable factor and for p_2 there exists a countable set $\{q_k\}$ of indecomposable probabilities such that

$$\lim_{k\to\infty} q_1 * \cdots * q_k \stackrel{c}{=} p_2 .$$

Proof Let p be a probability and m be a positive constant such that $p \in \mathcal{Q}_m$. If p has no indecomposable factor, the theorem is trivially proved. Otherwise, let

$$\eta_1 = \sup_{h \in K_1} H_m(h) > 0$$

where K_1 is the set of all the indecomposable factors of p and $q_1 \in K_1$ be such that

$$H_m(q_1) > \tfrac{1}{2}\eta_1.$$

Then

$$p = q_1 * r_1$$

where r_1 is a probability.

If r_1 has no indecomposable factor, the theorem is proved. Otherwise, let

$$\eta_2 = \sup_{h \in K_2} H_m(h) > 0$$

where K_2 is the set of all the indecomposable factors of r_1 and $q_2 \in K_2$ be such that

$$H_m(q_2) > \tfrac{1}{2}\eta_2 .$$

Then

$$p = q_1 * q_2 * r_2$$

where r_2 is a probability and so on.

If there exists some k such that r_k has no indecomposable factor, the theorem is proved. It remains to prove the theorem when we can continue indefinitely the decomposition process. We have then an infinite sequence $\{q_j\}$ of indecomposable probabilities such that

$$\sum_{j=1}^{k} H_m(q_j) \leq H_m(p)$$

for any integer k. It follows then that $\sum_{j=1}^{\infty} H_m(q_j)$ converges, so that

$$\lim_{k\to\infty} \sum_{j=k}^{k+l} H_m(q_j) = 0$$

uniformly with respect to l. From Theorem 4.5.1, there exists $\alpha_{k,l} \in R^n$ such that

$$\lim_{k\to\infty} e^{i(t,\alpha_{k,l})} \prod_{j=k}^{k+l} \hat{q}_j(t) = 1 \tag{4.5.7}$$

uniformly with respect to l and to t belonging to a compact set of R^n. Letting

$$\hat{q}_j(t) = \rho_j(t)e^{i\omega_j(t)}$$

where $\rho_j > 0$ and ω_j is continuous, we obtain from (4.5.7) the relation

$$(t, \alpha_{k,l}) + \sum_{j=k}^{k+l} \omega_j(t) = 2\pi b_{k,l}(t) + o(1) \qquad (k \to +\infty) \tag{4.5.8}$$

where $b_{k,l}$ is an integer. Since the left side of (4.5.8) is continuous and $b_{k,l}(0) = 0$, we have

$$b_{k,l}(t) = 0$$

if k is great enough. Therefore

$$(t, \alpha_{k,l}) + \sum_{j=k}^{k+l} \omega_j(t) = o(1) \qquad (k \to +\infty). \tag{4.5.9}$$

Since we can always assume (by multiplying $\hat{q}_j$ by a degenerate factor) that

$$\omega_j(e_\nu) = 0 \qquad (\nu = 1, \ldots, n),$$

we have

$$(t, \alpha_{k,l}) = o(1) \qquad (k \to +\infty)$$

and this implies, with (4.5.9),

$$\sum_{j=k}^{k+l} \omega_j(t) = o(1) \qquad (k \to +\infty),$$

so that

$$\lim_{k\to\infty} \prod_{j=k}^{k+l} \hat{q}_j(t) = 1 \tag{4.5.10}$$

uniformly with respect to l and t belonging to a compact set of R^n. Therefore $\prod_{j=1}^{\infty} \hat{q}_j(t)$ converges uniformly in any compact set of R^n, so that

$$\lim_{k\to\infty} (q_1 * \cdots * q_k) \overset{c}{=} p_2 .$$

If we consider now the sequence $\{r_k\}$, we have, by definition

$$r_k = q_{k+1} * \cdots * q_{k+l} * r_{k+l},$$

so that

$$\hat{r}_k(t) = \prod_{j=k+1}^{k+l} \hat{q}_j(t)\hat{r}_{k+l}(t)$$

and, from (4.5.10), we have

$$\lim_{k\to\infty} (\hat{r}_k(t) - \hat{r}_{k+l}(t)) = 0$$

uniformly with respect to l and t belonging to any compact set of R^n, so that

$$r_k \overset{c}{\to} p_1$$

and, by definition,

$$p = p_1 * p_2 .$$

It remains to prove that p_1 has no indecomposable factor. If p_1 has an indecomposable factor q, then, by definition, q belongs to K_k for any integer k so that

$$H_m(q) \leq \eta_k$$

for any k. Since

$$\eta_k < 2H_m(q_k) \qquad \text{and} \qquad \lim_{k\to\infty} H_m(q_k) = 0,$$

we have

$$H_m(q) = 0$$

and, from (4.5.4), q is degenerate.

This decomposition is not unique [see Lukacs (1970, p. 176)].

4.6 PROBABILFTIES WITH NO INDECOMPOSABLE FACTOR

Theorem 4.5.2 introduces a new class of probabilities (which we denote by I_0^n): the class of all the n-variate probabilities which have no indecomposable factor. We study in this section the simplest properties of this class. For this we introduce a new functional c_m which measures to some extent the amount of divisibility of a probability.

Let $m > 0$. If $p \in \mathcal{Q}_m$, for each decomposition D of p

$$p = p_1 * \cdots * p_{l_D},$$

we define the number

$$\gamma(D) = \sup_{1 \leq k \leq l_D} H_m(p_k)$$

and then

$$c_m(p) = \inf_D \gamma(D)$$

where the infimum is taken for all the decompositions of p. We have evidently

$$0 \leq c_m(p) \leq H_m(p)$$

and $c_m(p) = H_m(p)$ if and only if p is indecomposable. We study now the opposite situation with

Lemma 4.6.1 $c_m(p) = 0$ if and only if p is infinitely divisible.

Proof *Necessity* Let $p \in I^n$. If D_k is the decomposition

$$p = p_k^{*k},$$

then

$$\gamma(D_k) = (1/k)H_m(p),$$

so that

$$c_m(p) = \inf_D \gamma(D) = 0.$$

Sufficiency If $c_m(p) = 0$, there exists a sequence of decompositions D_j

$$p = p_{j,1} * \cdots * p_{j,l_j}$$

such that

$$\lim_{j\to\infty} \sup_{1 \le k \le l_j} H_m(p_{j,k}) = 0.$$

From Theorem 4.5.1, there exists $\alpha_{j,k} \in R^n$ such that

$$\lim_{j\to\infty} \sup_{1 \le k \le l_j} p_{j,k}(\{x : \|x - \alpha_{j,k}\| > \varepsilon\}) = 0$$

for any $\varepsilon > 0$. If

$$q_{j,k} = p_{j,k} * \delta_{\alpha_{j,k}},$$

then

$$\lim_{j\to\infty} \sup_{1 \le k \le l_j} q_{j,k}(\{x : \|x\| > \varepsilon\}) = 0$$

for any $\varepsilon > 0$ and $\{q_{j,k}\}$ is a null array. Since

$$p = q_{j,1} * \cdots * q_{j,l_j} * \delta_{-\alpha_j}$$

where

$$\alpha_j = \sum_{k=1}^{l_j} \alpha_{j,k},$$

p is the limit of the array $\{q_{j,k}\}$ for the centering $\{-\alpha_j\}$ and, from Theorem 4.4.2, we deduce that p is infinitely divisible.

We prove now

Lemma 4.6.2 If $p \in I_0^n$, then $c_m(p) = 0$ for some positive constant m.

Proof By definition of $c_m(p)$, there exists a sequence of decompositions D_j of p

$$p = p_{j,1} * \cdots * p_{j,l_j}$$

such that

$$H_m(p_{j,1}) \leq \cdots \leq H_m(p_{j,l_j})$$

and

$$c_m(p) \leq \gamma(D_j) \leq c_m(p) + (1/j) \qquad (j = 1, 2, \ldots).$$

Let q_j and r_j be defined by

$$p_{j,1} = q_j, \qquad p_{j,2} * \cdots * p_{j,l_j} = r_j.$$

According to Helly's first theorem, the sequence $\{q_j\}$ contains a subsequence $\{q_{j_k}\}$ which converges weakly to a measure q and, from Theorem 4.5.1, q is a probability. Likewise, from $\{r_{j_k}\}$ we can extract a subsequence which converges completely to a probability r. Then

$$p = q * r \qquad \text{and} \qquad H_m(q) = c_m(p).$$

If we suppose that

$$c_m(p) \geq \tfrac{1}{2}H_m(p),$$

then

$$H_m(r) \leq c_m(p),$$

and since q and r are both decomposable, we have a decomposition D_1' with $\gamma(D_1') < c_m(p)$ and that contradicts the definition of $c_m(p)$. We have then a decomposition D_1' of p such that

$$\gamma(D_1') < \tfrac{1}{2}H_m(p).$$

If we apply this result to D_1', we have a decomposition D_2' such that

$$\gamma(D_2') < (1/2^2)H_m(p)$$

and, more generally, a decomposition D_k' such that

$$\gamma(D_k') < (1/2^k)H_m(p).$$

Since

$$c_m(p) \leq (D_k')$$

for any k, we have

$$c_m(p) = 0.$$

From these two lemmas, we deduce immediately the following theorem.

Theorem 4.6.1 An n-variate probability which has no indecomposable factor is infinitely divisible (in other words, $I_0^n \subset I^n$).

Corollary 4.6.1 If $p \in I_0^n$ and if q divides p, then q is infinitely divisible.

From this result and Corollary 4.3.7, we deduce immediately

Theorem 4.6.2 If p belongs to I_0^n, then $\hat{p}$ admits an (α, P, μ)-Lévy–Hinčin representation and q divides p if and only if $\hat{q}$ admits a (β, Q, ν)-Lévy–Hinčin representation with

$$0 \leq Q \leq P, \qquad 0 \leq \nu \leq \mu.$$

Corollary 4.6.2 If p belongs to I_0^n and if q divides p, then the Poisson spectrum of p contains the Poisson spectrum of q.

Corollary 4.6.3 If p belongs to I_0^n and if $\hat{p}$ admits an (α, P, μ)-De Finetti representation, then q divides p if and only if $\hat{q}$ admits a (β, Q, ν)-De Finetti representation with

$$0 \leq Q \leq P, \qquad 0 \leq \nu \leq \mu.$$

We prove now that I_0^n contains the class of all the normal probabilities with

Theorem 4.6.3 (*Cramér's theorem*) Any factor of an n-variate normal probability is an n-variate normal probability.

Proof In the case $n = 1$, this result is well known [see, for example, Lukacs (1970, Theorem 8.2.1)]. The general case follows from this case and Theorem 3.3.5.

Finally, we mention the problem of the description of the class I_0^n. If $\hat{p}$ admits an (α, P, μ)-Lévy–Hinčin representation with a finite measure μ, what condition on P and μ implies that p belongs to I_0^n?

This problem is not yet solved and we will give its present status in the following chapters.

NOTES

The notion of indecomposable probability is due to Hinčin (1937b) and Lévy (1937a, 1938a). The density of the set of indecomposable probabilities was proved by Parthasarathy *et al.* (1962) in a more general context. Theorems 4.1.2 and 4.1.3 are taken from Linnik and Ostrovskiĭ (1972).

The first work on infinitely divisible probabilities is due to De Finetti (1929) who proved Theorem 4.2.4 and studied the characteristic functions (4.3.1) for an univariate measure μ. Then Kolmogorov (1932) characterized the univariate infinitely divisible probabilities having a finite moment of order 2 and Lévy (1934) characterized the univariate infinitely divisible probabilities, another proof being due to Hinčin (1937a). Lévy (1937b) solved also the multivariate case. Theorem 4.2.6 was proved by Schoenberg (1938) in another context. The limit theorems 4.3.5 and 4.4.2 are due to Gnedenko (1939) in the case $n = 1$ and to Rvačeva (1954) in the general case. Theorem 4.3.7 is due to Cuppens (1969g) and Theorems 4.3.8 and 4.3.9 are from Ostrovskiĭ (1970).

Theorems 4.5.2 and 4.6.1 are due to Hinčin (1937b). The link between these theorems and the existence of the functional H_m and Theorem 4.4.2 has been studied by Kendall (1968) and Davidson (1968, 1969). Theorem 4.6.3 was conjectured by Lévy (1934) and proved by Cramér (1936).

Chapter 5

Decomposition Theorems for Analytic Characteristic Functions

5.1 DECOMPOSITIONS OF DERIVABLE CHARACTERISTIC FUNCTIONS

Theorem 5.1.1 If an n-variate probability p has a moment of order k' for any $k' \leq k$ ($k \in N^n$) and if q divides p, then q has a moment of order k.

Proof We suppose that the theorem has been proved for any $k' \leq k$, $k' \neq k$ and that

$$\beta_k = \int |u|^k p(du) < +\infty.$$

If $p = q * r$, then, from Theorem 2.5.3, we have

$$\beta_k = \int \left(\int |y + z|^k q(dz) \right) r(dy).$$

Let $m \in R^n$ be such that $r(C_m) > 0$ where

$$C_m = \{x \in R^n : -m \leq x \leq m\}.$$

Then

$$\beta_k \geq \int_{C_m} \left(\int_{C_m{}^c} | y + z |^k q(dz) \right) r(dy)$$

$$\geq r(C_m) \int_{C_m{}^c} (| z | - | m |)^k q(dz) = r(C_m) \left(\int_{C_m{}^c} |z|^k q(dz) + S \right)$$

where S is a sum of integrals of the kind

$$\int_{C_m{}^c} | z |^{k'} q(dz)$$

with $k' \leq k$, $k' \neq k$. Since these integrals converge, we obtain the existence of

$$\int_{C_m{}^c} | z |^k q(dz)$$

and therefore the existence of the moment of order k of q since

$$\int_{C_m} | z |^k q(dz) \leq | m |^k.$$

The converse which is true when $n = 1$ is false in the general case. For example, if $n = 2$ and if $p = q \otimes r$ where q is concentrated on the first coordinate axis and r is concentrated on the second coordinate axis, then

$$\int u_1 u_2 q(du) = \int u_1 u_2 r(du) = 0,$$

but $\int u_1 u_2 p(du)$ exists if and only if $\int u_1 q(du)$ and $\int u_2 r(du)$ exist.

From Theorems 5.1.1 and 3.2.4 and Corollary 3.2.2, we have

Corollary 5.1.1 If, for some $k \in N^n$, $D^{2k}\hat{p}(0)$ exists and if q divides p, then $D^{2k}\hat{q}(0)$ exists.

This result is not true for derivatives which are not of even order. An example in the case $n = 1$ has been given by Wolfe (1971).

Theorem 5.1.2 If $\hat{p}$ is an infinitely divisible characteristic function admitting an (α, P, μ)-Lévy–Hinčin representation, then $D^{2k}\hat{p}(0)$ exists for some $k \in N^n$ if and only if $D^{2k}\hat{\mu}(0)$ exists.

Proof *Necessity* We consider the restriction ν of μ to $A = \{x \in R^n: \|x\| > 1\}$ such that

$$\nu(B) = \mu(B \cap A)$$

for any $B \in \mathscr{B}_n$. Since for $\nu = 0$, the theorem is clear, we can suppose that $\nu \neq 0$. From Corollary 4.3.7, we deduce that the probability q defined by

$$\log \hat{q}(t) = \int K(t, u)\nu(du)$$

divides p. But we easily see that $\exp \nu/\exp\|\nu\|$ divides q and, from Corollary 5.1.1, $D^{2k} \exp(\hat{\nu}(0))$ exists. This implies the existence of $D^{2k}\hat{\nu}(0)$ and, from Corollary 3.2.2, the existence of all the moments of $\nu/\|\nu\|$ of order $k' \leq 2k$. From this, we deduce the existence of all the moments of $\mu/\|\mu\|$ of order $k' \leq 2k$ and, from Theorem 3.2.4, the existence of $D^{2k}\hat{\mu}(0)$.

Sufficiency Since for $\mu = 0$, the theorem is clear, we can suppose that $\mu \neq 0$. If $D^{2k}\hat{\mu}(0)$ exists, from Corollary 3.2.2, we deduce the existence of all the moments of μ of order $k' \leq 2k$. Defining ν' by

$$\nu'(B) = \int_{B \cap A} \frac{1 + \|u\|^2}{\|u\|^2} \mu(du)$$

for any $B \in \mathscr{B}_n$, we have

$$\nu' \leq 2\mu$$

and the moments of ν' of order $k' \leq 2k$ exist. From theorem 3.2.4, $D^{2k}\hat{\nu}'(0)$ exists and therefore $D^{2k} \exp(\hat{\nu}'(0))$ exists. If $\hat{q}$ is the characteristic function

$$\hat{q}(t) = \exp\left(\int_{A^c} K(t, u)\mu(du)\right),$$

we deduce from Theorem 3.6.5 that $\hat{q}$ is an entire function. Since

$$\hat{p} = \hat{n}(\alpha, P) \cdot \hat{q} \cdot \frac{\exp(\nu')}{\exp(\|\nu'\|)},$$

$\hat{p}$ is the product of three functions having a derivative of order $2k$ at the origin and therefore $D^{2k}\hat{p}(0)$ exists.

5.2 DECOMPOSITIONS OF PROBABILITIES BELONGING TO $\mathscr{A}_\Gamma$

Theorem 5.2.1 If $p = q * r$, then $\int e^{(y, u)}p(du)$ exists for some $y \in R^n$ if and only if $\int e^{(y, u)}q(du)$ and $\int e^{(y, u)}r(du)$ both exist.

Proof If $p = q * r$, then, from Theorem 2.5.3, $\int e^{(y, u)}p(du)$ exists if and only if $\int (\int e^{(y, u+v)}q(dv))r(du)$ exists. Since

$$\int \left(\int e^{(y, u+v)}q(dv)\right)r(du) = \int e^{(y, v)}q(dv) \int e^{(y, u)}r(du),$$

we easily obtain the theorem.

Corollary 5.2.1 Let p be an n-variate probability belonging to $\mathscr{A}_\Gamma$ for some convex set $\Gamma \subset R^n$ containing the origin. If

$$\hat{p}(t) = \hat{q}(t)\hat{r}(t) \tag{5.2.1}$$

for any $t \in R^n$, then q and r belong to $\mathscr{A}_\Gamma$ and (5.2.1) holds for any $t \in R^n + i\Gamma$. We have then

$$|\hat{p}(t + iy)/\hat{p}(iy)| \le |\hat{q}(t + iy)/\hat{q}(iy)| \le 1 \tag{5.2.2}$$

for any $t \in R^n$, $y \in \Gamma$ and there exist some $C > 0$ and $m \in R^n$ such that

$$\hat{q}(iy) \le Ce^{(m, |y|)}\hat{p}(iy)$$

for any $y \in \Gamma$.

Proof From Theorem 5.2.1, q and r belong to $\mathscr{A}_\Gamma$ and, from Theorem 2.5.3,

$$\begin{aligned}\hat{p}(t) &= \int e^{i(t, u)}p(du) = \int \left(\int e^{i(t, u+v)}q(dv)\right)r(du) \\ &= \int e^{i(t, v)}q(dv) \int e^{i(t, u)}r(du) = \hat{q}(t)\hat{r}(t)\end{aligned}$$

for any $t \in R^n + i\Gamma$. Since $\hat{q}$ and $\hat{p}/\hat{q}$ are characteristic functions of probabilities belonging to $\mathscr{A}_\Gamma$, (5.2.2) follows easily from the ridge property. The last assertion follows from Corollary 3.5.3.

From Corollaries 5.2.1 and 3.5.2, we obtain

Corollary 5.2.2 Let p be a probability belonging to $\mathscr{A}_\Gamma$ and p_y be the probability defined by

$$\hat{p}_y(t) = \hat{p}(t + iy)/\hat{p}(iy)$$

for any $y \in \Gamma$. If q divides p, then the probability q_y defined by

$$\hat{q}_y(t) = \hat{q}(t + iy)/\hat{q}(iy)$$

divides p_y.

Corollary 5.2.3 Let q be a factor of $p \in \mathscr{A}_\Gamma$. If $\hat{p}(t) \ne 0$ for some $t \in R^n + i\Gamma$, then $\hat{q}(t) \ne 0$. If $\hat{p}$ has no zeros in $R^n + i\Gamma$ and if ρ and σ are defined by

$$\rho(x, y) = \mathrm{Re}(\log \hat{p}(iy) - \log \hat{p}(x + iy)),$$

$$\sigma(x, y) = \mathrm{Re}(\log \hat{q}(iy) - \log \hat{q}(x + iy)),$$

then

$$0 \le \sigma(x, y) \le \rho(x, y)$$

for any $x \in R^n$, $y \in \Gamma$.

Proof The first assertion follows directly from Corollary 5.2.1 and the second can be deduced easily from (5.2.2).

From Corollary 5.2.1 and Theorem 3.6.3, we obtain easily

Corollary 5.2.4 If an n-variate probability p is bounded in a direction θ and if $p = q * r$, then q and r are bounded in the direction θ and

$$\text{ext}_\theta\, p = \text{ext}_\theta\, q + \text{ext}_\theta\, r.$$

We give now a more specialized result.

Theorem 5.2.2 Let $\Gamma = \{x \in R^n : x = l\theta, l \geq 0\}$ $(\theta \in R^n, \|\theta\| = 1)$ and p be a probability belonging to $\mathscr{A}_\Gamma$. If $\hat{p}$ has no zeros in $R^n + i\Gamma$ and if q divides p, there exists some positive constant C such that

$$|\log \hat{q}(k\theta + \phi)| \leq (2|k|/\eta)\Bigg(2 \sup_{0 \leq u \leq \text{Im}\, k + \eta} |\log \hat{p}(iu\theta)| + \sup_{0 \leq \text{Im}\, u \leq \text{Im}\, k + \eta} |\log \hat{p}(u\theta + \phi)| + C(\text{Im}\, k + \eta)\Bigg)$$

for any k satisfying $\text{Im}\, k > \eta > 0$ and any ϕ orthogonal to θ.

Proof From Corollary 5.2.2, we deduce that $\hat{q}$ is the characteristic function of a probability belonging to $\mathscr{A}_\Gamma$, has no zeros in $R^n + i\Gamma$ and satisfies

$$0 \leq \sigma_0(0, l) - \sigma_\phi(k, l) \leq \rho_0(0, l) - \rho_\phi(k, l) \tag{5.2.3}$$

where

$$\rho_\phi(k, l) = \text{Re} \log \hat{p}((k + il)\theta + \phi),$$
$$\sigma_\phi(k, l) = \text{Re} \log \hat{q}((k + il)\theta + \phi)$$

$(\phi \perp \theta, k \in R, l \geq 0)$. From Corollaries 5.2.3 and 3.5.3 and the relations

$$\rho_0(0, l) = \log \hat{p}(il\theta), \qquad \sigma_0(0, l) = \log \hat{q}(il\theta),$$

we obtain

$$\sigma_0(0, l) \geq -m_1 - m_2 l, \qquad \sigma_0(0, l) \leq \rho_0(0, l) + m_3 + m_4 l$$

where $m_j \geq 0$ $(j = 1, 2, 3, 4)$, so that

$$|\sigma_0(0, l)| \leq |\rho_0(0, l)| + C_1 l + C_2 \tag{5.2.4}$$

where $C_j \geq 0$ $(j = 1, 2)$. From (5.2.3) and (5.2.4), we obtain

$$|\sigma_\phi(k, l)| \leq 2|\rho_0(0, l)| + |\rho_\phi(k, l)| + C_1 l + C_2 .$$

If

$$f_\phi(z) = \log \hat{p}(z\theta + i\phi), \qquad g_\phi(z) = \log \hat{q}(z\theta + i\phi),$$

we have for $k > \eta$ (see Appendix A.4)

$$g'_\phi(k + il) = (1/\pi\eta) \int_0^{2\pi} \sigma_\phi(k + \eta \cos u, l + \eta \sin u) e^{-iu}\, du,$$

so that

$$|g'_\phi(v)| \le (2/\eta)\Bigg(2 \sup_{\operatorname{Im} v - \eta \le u \le \operatorname{Im} v + \eta} |f_0(iu)| + \sup_{|v - z| = \eta} |f_\phi(z)| + C_1(\operatorname{Im} v + \eta) + C_2\Bigg).$$

Since, for any $\eta' > \eta$

$$g_\phi(z) = \int_{i\eta'}^{z} g'_\phi(v)\, dv + g_\phi(i\eta'),$$

we have for $\operatorname{Im} z > \eta$

$$|g_\phi(z)| \le (2/\eta)|z|\Bigg(2 \sup_{u \le \operatorname{Im} z + \eta} |f_0(iu)| + \sup_{\operatorname{Im} u \le \operatorname{Im} z + \eta} |f_\phi(u)| + C_1(\operatorname{Im} z + \eta) + C_2\Bigg)$$

and from this we easily deduce the theorem.

Corollary 5.2.5 Let $\hat{p}$ and $\hat{q}$ be two characteristic functions admitting an $(\alpha, 0, \mu)$-De Finetti representation and a $(\beta, 0, \nu)$-De Finetti representation, respectively. If q divides p and if μ is bounded in the direction θ, then ν is bounded in the direction θ and

$$\operatorname{ext}_\theta \nu \le \sup(0, \operatorname{ext}_\theta \mu).$$

Proof From Theorem 5.2.2, we deduce that

$$\sup_{\phi \perp \theta} \limsup_{r \to \infty} \frac{\log |g_\phi(ir)|}{r} \le \sup\left(\sup_{\phi \perp \theta} \limsup_{r \to \infty} \frac{\log |f_\phi(ir)|}{r}, 0\right)$$

where $f_\phi(k) = \log \hat{p}(k\theta + \phi)$, $g_\phi(k) = \log \hat{q}(k\theta + \phi)$. From this relation and Theorem 3.6.1, we deduce the corollary.

For infinitely divisible probabilities, we have also the following results.

Theorem 5.2.3 If an infinitely divisible characteristic function $\hat{p}$ admits an (α, P, μ)-Lévy–Hinčin representation

$$\log \hat{p}(t) = i(\alpha, t) - P(t) + \int K(t, u)\mu(du) \tag{5.2.5}$$

and if p belongs to $\mathscr{A}_\Gamma$, then $\hat{p}$ has no zeros in $R^n + i\Gamma$ and representation (5.2.5) holds for any $t \in R^n + i\Gamma$.

Proof For the case $n = 1$, see Lukacs (1970, Theorem 11.2.1). The general case follows from the case $n = 1$ and Theorem 4.3.2.

From the corresponding result in the case $n = 1$ [see Lukacs (1970, Theorem 11.2.2)], we obtain also

Theorem 5.2.4 If an infinitely divisible characteristic function $\hat{p}$ admits an (α, P, μ)-Lévy–Hinčin representation, p is bounded in the direction θ if and only if the following conditions are satisfied:

(a) $P(\theta) = 0$;
(b) μ is concentrated on the half-space $\{x \in R^n : (\theta, x) \leq 0\}$;
(c) $\int [(\theta, u)/\|u\|^2]\mu(du)$ exists.

Then

$$\text{ext}_\theta\, p = (\alpha, \theta) - \int [(\theta, u)/\|u\|^2]\mu(du).$$

With the method used in the proof of Theorem 5.1.2, we obtain also

Theorem 5.2.5 If the infinitely divisible characteristic function $\hat{p}$ admits an (α, P, μ)-Lévy–Hinčin representation, then $\hat{p}$ belongs to $\mathscr{A}_\Gamma$ if and only if $\hat{\mu}/\|\mu\|$ belongs to $\mathscr{A}_\Gamma$.

5.3 DECOMPOSITIONS OF ANALYTIC CHARACTERISTIC FUNCTIONS

All the results of Section 5.2 hold for analytic characteristic functions. Nevertheless, we have a more precise result.

Theorem 5.3.1 Let $\hat{p}$ be an n-variate characteristic function which is regular in the convex tube $R^n + i\Gamma$. If there exists some $\rho > 0$ such that

$$\hat{p}(t) = \hat{q}(t)\hat{r}(t) \tag{5.3.1}$$

holds for any $t \in R^n$, $\|t\| \leq \rho$, then $\hat{q}$ and $\hat{r}$ are regular in $R^n + i\Gamma$ and (5.3.1) is valid for any $t \in R^n + i\Gamma$. Moreover, there exist some positive constant C and some $m \in R^n$ such that

$$\hat{q}(iy) \leq Ce^{(m, |y|)}\hat{p}(iy)$$

for any $y \in \Gamma$.

Proof When $n = 1$, this result is a well-known theorem due to Raĭkov [see Lukacs (1970, Theorem 8.1.1)]. In the general case, if $\theta \in R^n \backslash \{0\}$, we obtain from Corollary 3.3.3 that $\mathrm{pr}_\theta\, \hat{p}$ is analytic in the tube $\{z \in C^n : -j(-\theta) < \mathrm{pr}_\theta \,\mathrm{Im}\, z < j(\theta)\}$ where j is the gauge of Γ. From Raĭkov's theorem, it follows that $\mathrm{pr}_\theta\, \hat{q}$ and $\mathrm{pr}_\theta\, \hat{r}$ have the same property and, since this is true for any $\theta \in R^n \backslash \{0\}$, we obtain from Corollary 3.3.3 the analyticity of $\hat{q}$ and $\hat{r}$ in the tube $R^n + i\Gamma$. The validity of (5.3.1) in $R^n + i\Gamma$ is a direct consequence of the uniqueness of a regular extension and the last assertion follows immediately from Corollary 5.2.1.

Corollary 5.3.1 With the conditions of Theorem 5.3.1,

(a) if $\hat{p}$ is an entire function, then $\hat{q}$ and $\hat{r}$ are entire functions;

(b) if $\hat{p}$ is an entire function with finite order ρ, then $\hat{q}$ and $\hat{r}$ are entire functions with finite order not greater than ρ;

(c) if $\hat{p}$ is an entire function with exponential type, then $\hat{q}$ and $\hat{r}$ are entire functions with exponential type.

Using Theorem 3.6.4, we can make precise the last assertion of the preceding corollary with

Corollary 5.3.2 In the conditions of Theorem 5.3.1, if p is concentrated on a bounded convex set A_0, there exist some bounded convex sets A_1 and A_2 such that q is concentrated on A_1 and r is concentrated on A_2. Moreover

$$h_0 = h_1 + h_2$$

where h_j is the support function of A_j $(j = 0, 1, 2)$.

We have also

Theorem 5.3.2 Let p be a probability having a compact support and such that

$$p = p_1 \otimes p_2 .$$

If q divides p, then

$$q = q_1 \otimes q_2$$

where q_j divides p_j $(j = 1, 2)$.

Proof This follows immediately from Corollary 5.3.2 and the following result.

If the entire functions f_1 and f_2 of exponential type satisfy

$$f_1(t_1, \ldots, t_n) f_2(t_1, \ldots, t_n) = g(t_1, \ldots, t_m) h(t_{m+1}, \ldots, t_n),$$

then

$$f_j(t_1, \ldots, t_n) = g_j(t_1, \ldots, t_m) h_j(t_{m+1}, \ldots, t_n) \qquad (j = 1, 2).$$

Proof See Lepetit (1973).

This result is not true if the support of p is not compact, for example, if p is a normal probability.

From Theorem 5.2.2, we obtain

Theorem 5.3.3 If $\hat{p}$ is an entire characteristic function without zeros and if q divides p, for any $\xi > \eta > 0$, there exists some positive constant C such that

$$\sup_{\|z\| \le \xi} |\log \hat{q}(z)| \le (6\xi/\eta) \sup_{\|z\| \le \xi+\eta} |\log \hat{p}(z)| + C\xi(\xi + \eta).$$

We prove now a result which is the analog of Theorem 5.3.2 for logarithm of characteristic functions.

Theorem 5.3.4 Let $\hat{p}$ be an entire characteristic function admitting an $(\alpha, 0, \mu)$-Lévy–Hinčin representation for any $t \in C^n$. If

$$\mu = \mu_1 \oplus \mu_2 ,$$

then $p_j = \exp \mu_j/\exp\|\mu_j\|$ is a probability $(j = 1, 2)$. If q divides p and if $\hat{q}$ admits for any $t \in C^n$ a $(\beta, 0, \nu)$-Lévy–Hinčin representation, then

$$\nu = \nu_1 \oplus \nu_2$$

and $q_j = \exp \nu_j/\exp\|\nu_j\|$ is a probability dividing p_j $(j = 1, 2)$.

Proof The first assertion follows directly from

$$\hat{p}_j(t) = \hat{p}(\mathrm{pr}_{E_j}\, t)$$

where E_j is the subspace containing the support of μ_j. If we define ρ and σ by

$$\rho(x, y) = \mathrm{Re}(\log \hat{p}(iy) - \log \hat{p}(x + iy)),$$

$$\sigma(x, y) = \mathrm{Re}(\log \hat{q}(iy) - \log \hat{q}(x + iy)),$$

from Corollary 5.2.3, we have

$$0 \le \sigma(x, y) \le \rho(x, y)$$

and we obtain easily

$$\rho(x, y) = \sum_{j=1}^{2} \int \exp(-(\mathrm{pr}_{E_j}\, y, u)) \sin^2(\tfrac{1}{2}(\mathrm{pr}_{E_j}\, x, u)) \frac{1 + \|u\|^2}{\|u\|^2} \mu_j(du).$$

If we decompose ν as

$$\nu = \nu_1 + \nu_2 + \nu'$$

where

$$\nu_j(B) = \nu(B \cap E_j) \qquad (j = 1, 2)$$

for any $B \in \mathscr{B}_n$, we have

$$\sigma(x, y) = \sum_{j=1}^{2} \int \exp(-(\mathrm{pr}_{E_j}\, y, u)) \sin^2(\tfrac{1}{2}(\mathrm{pr}_{E_j}\, x, u)) \frac{1 + \|u\|^2}{\|u\|^2} \nu_j(du)$$

$$+ \int e^{-(y, u)} \sin^2(\tfrac{1}{2}(x, u)) \frac{1 + \|u\|^2}{\|u\|^2} \nu'(du).$$

Letting $\mathrm{pr}_{E_2}\, x = 0$, we obtain

$$0 \leq \int \exp(-(\mathrm{pr}_{E_1}\, y, u)) \sin^2(\tfrac{1}{2}(\mathrm{pr}_{E_1}\, x, u)) \frac{1 + \|u\|^2}{\|u\|^2} \nu_1(du)$$

$$+ \int e^{-(y, u)} \sin^2(\tfrac{1}{2}(\mathrm{pr}_{E_1}\, x, u)) \frac{1 + \|u\|^2}{\|u\|^2} \nu'(du)$$

$$\leq \int \exp(-(\mathrm{pr}_{E_1}\, y, u)) \sin^2(\tfrac{1}{2}(\mathrm{pr}_{E_1}\, x, u)) \frac{1 + \|u\|^2}{\|u\|^2} \mu_1(du).$$

From this we obtain easily

$$\int e^{-(y, u)} \sin^2(\tfrac{1}{2}(\mathrm{pr}_{E_1}\, x, u)) \frac{1 + \|u\|^2}{\|u\|^2} \nu'(du) = 0$$

and this implies

$$\nu_x(B) = \int_B \sin^2(\tfrac{1}{2}(\mathrm{pr}_{E_1}\, x, u)) \frac{1 + \|u\|^2}{\|u\|^2} \nu'(du) = 0$$

for any $B \in \mathscr{B}_n$, so that ν' is concentrated on

$$\bigcap_x \{u : \sin^2(\tfrac{1}{2}(\mathrm{pr}_{E_1}\, x, u)) = 0\} = \{u : \mathrm{pr}_{E_2}\, u = 0\} = E_1$$

and this implies $\nu' = 0$ and

$$\nu = \nu_1 \oplus \nu_2 .$$

The last assertion of the theorem is evident.

Finally, we prove

Theorem 5.3.5 (a) Any factor of an n-variate Poisson-type probability is a Poisson-type probability.

(b) Any factor of the convolution of an n-variate normal probability and an n-variate Poisson-type probability is the convolution of a normal probability and a Poisson-type probability.

Proof In the case $n = 1$, (a) is the well-known Raĭkov's theorem [see Lukacs (1970, Theorem 8.2.2)] and (b) is a theorem due to Linnik [see

Lukacs (1970, Theorem 8.2.3)]. We obtain the general case from the case $n = 1$ and Theorems 5.3.1, 3.4.3, and 3.4.4.

From this theorem, we deduce easily that the convolution of a normal probability and a Poisson-type probability belongs to I_0^n.

NOTES

For univariate probabilities and strips which are symmetric with respect to the real axis, Theorem 5.3.1 is due to Raĭkov (1938a) and Theorem 5.2.3 to Raĭkov (1938b). Wintner (1938) proved Theorems 5.1.1 and 5.2.1 for the case $n = 1$.

Theorems on multivariate analytic characteristic functions were proved independently by Ostrovskiĭ (1966a) and Cuppens (1967). The extensions to $\mathscr{A}_\Gamma$ are due to Cuppens (1967).

Theorem 5.3.5(a) is due to Raĭkov (1937) for the case $n = 1$ and to Teicher (1954) for the general case, while Theorem 5.3.5(b) was proved by Linnik (1957) and more simply by Ostrovskiĭ (1965b) for the case $n = 1$ and independently by Ostrovskiĭ (1965c) and Cuppens (1967) for the general case.

Chapter 6

Infinitely Divisible Probabilities with Normal Factor

We study now the problem stated at the end of Chapter 4 for the description of the class I_0^n and begin with the class of infinitely divisible probabilities with normal factors. We introduce here a new terminology which is simpler than the one used in the case $n = 1$ by Lukacs (1970).

6.1 CASE $n = 1$

We introduce the following definition. A set Λ of real numbers is a *Linnik set* if it satisfies the following condition: for any $(a, b) \in \Lambda^2$ such that $\operatorname{sgn} a = \operatorname{sgn} b$, then either ab^{-1} or ba^{-1} is an integer greater than 1.

From this definition, it follows that a Linnik set is countable and has only 0 as a possible accumulation point. For convenience, we suppose that 0 belongs always to a Linnik set and, if $\Lambda = \{\lambda_k\}$, we suppose that $\lambda_0 = 0$.

With this definition, we have the following results.

Theorem 6.1.1 If a univariate probability has a nondegenerate normal factor and no indecomposable factor, then its Poisson measure is concentrated on a Linnik set.

Proof See Linnik (1960, Chapter 8).

Theorem 6.1.2 Let $\hat{p}$ be a univariate characteristic function admitting an (α, P, μ)-Lévy–Hinčin representation. If μ is concentrated on a Linnik set and satisfies

$$\mu(\{x : \|x\| \geq u\}) = O(\exp(-Ku^2)) \qquad (u \to +\infty) \tag{6.1.1}$$

for some $K > 0$, then p has no indecomposable factor.

Proof See Lukacs (1970, Theorem 9.2.1).

It follows from Theorem 4.6.2 and Corollary 4.6.2 that this result is equivalent to

Corollary 6.1.1 Let $\hat{p}$ be a univariate infinitely divisible characteristic function admitting an (α, P, μ)-Lévy–Hinčin representation where μ is concentrated on a Linnik set Λ and satisfies (6.1.1). Then q divides p if and only if $\hat{q}$ admits a (β, Q, ν)-Lévy–Hinčin representation with

$$0 \leq Q \leq P, \qquad 0 \leq \nu \leq \mu.$$

Moreover, ν is concentrated on Λ and satisfies

$$\nu(\{x : \|x\| \geq u\}) = O(\exp(-Ku^2)) \qquad (u \to +\infty).$$

We want to extend these results to the case $n > 1$.

6.2 A NECESSARY CONDITION

A set $\Lambda \subset R^n$ is a Linnik set constructed on $\{\theta_1, \ldots, \theta_n\}$ if

$$\Lambda = \Lambda_1 + \cdots + \Lambda_n$$

where $\Lambda_j = \{\lambda_{j,m}\theta_j\}$ $(j = 1, \ldots, n)$, $\{\theta_1, \ldots, \theta_n\}$ being a basis of R^n and $\{\lambda_{j,m}\}$ being a Linnik set of real numbers $(j = 1, \ldots, n)$.

We do not know if a necessary condition similar to the one given by Theorem 6.1.1 for univariate probabilities holds for n-variate probabilities. Nevertheless, we have

Theorem 6.2.1 Let $\hat{p}$ be an n-variate characteristic function admitting an (α, P, μ)-Lévy–Hinčin representation. If p belongs to I_0^n and if $\theta \neq 0$ belongs to the support of the normal factor of p, then the restriction $t_\theta \mu$ of μ to the one-dimensional subspace generated by θ is concentrated on a Linnik set $\Lambda = \{\lambda_m \theta\}$.

Proof Let $\hat{q}$ be the characteristic function admitting a $(0, Q, t_\theta \mu)$-Lévy–Hincin representation where Q is a quadratic form such that

$$Q = \mathrm{pr}_\theta\, Q$$

and $P - Q$ is a nonnegative quadratic form. (The existence of such a quadratic form follows immediately from the results on the decompositions in squares of quadratic forms.) Then, from Corollary 4.3.7, q divides p and if $t_\theta \mu$ is not concentrated on a Linnik set, it follows from Theorem 6.1.1 that q has an indecomposable factor, so that p does not belong to I_0^n.

We must note that this necessary condition is not very satisfactory since it is satisfied by many continuous measures μ while no univariate continuous measure satisfies the condition of Theorem 6.1.1.

6.3 INDUCTION METHOD

We begin this section with some lemmas.

Lemma 6.3.1 Let $\hat{p}$ and $\hat{q}$ be two entire characteristic functions. We suppose that $\hat{p}$ admits for any $z \in C^n$ a De Finetti representation

$$\log \hat{p}(z) = i(\alpha, z) - P(z) + \int (e^{i(z, u)} - 1)\mu(du)$$

where $\alpha \in R^n$, $\mu \in \mathscr{M}_n$ and P is a nonnegative quadratic form, and that $\hat{q}$ admits for any $z \in C^n$ the representation

$$\log \hat{q}(z) = \Phi(z) + \Psi(z)$$

where Φ is a polynomial with degree τ

$$\Phi(z) = \sum_k c_k (iz)^k,$$

c_k being real constants and Ψ being a polynomial with degree τ

$$\Psi(z) = \sum_k \left(\int e^{i(z, u)} \nu_k(du) \right) (iz)^k,$$

the coefficients being Fourier–Stieltjes transforms of finite signed measures ν_k. If q divides p, then

$$\log \hat{q}(z) = i(\beta, z) - Q(z) + \int (e^{i(z, u)} - 1)\nu_0(du) + i \sum_{j=1}^{n} \left(\int e^{i(z, u)} \nu_j(du) \right) z_j \tag{6.3.1}$$

where $\beta \in R^n$, $\nu_j \in \mathscr{M}_n$ $(j = 0, 1, \ldots, n)$ and Q is a quadratic form satisfying

$$0 \le Q(z) \le P(z)$$

for any $z \in R^n$.

Proof We prove this result by induction on τ and begin with the case $\tau = 2$. If we define ρ and σ by

$$\rho(x, y) = \operatorname{Re}(\log \hat{p}(iy) - \log \hat{p}(x + iy)),$$
$$\sigma(x, y) = \operatorname{Re}(\log \hat{q}(iy) - \log \hat{q}(x + iy))$$

for any $x \in R^n$, $y \in R^n$, we obtain, from Corollary 3.3.2,

$$0 \le \sigma(x, y) \le \rho(x, y) \tag{6.3.2}$$

and we find easily that

$$\rho(x, y) = P(x) + \int e^{-(y,u)} \sin^2(\tfrac{1}{2}(x, u))\mu(du) \tag{6.3.3}$$

and

$$\begin{aligned}\sigma(x, y) = &\int e^{-(y,u)} \sin^2(\tfrac{1}{2}(x, u))\nu_0(du)\\ &+ \sum_{j=1}^{n} \int e^{-(y,u)}[x_j \sin(x, u) - y_j \sin^2(\tfrac{1}{2}(x, u))]\nu_{e_j}(du)\\ &+ \sum_{j=1}^{n} \sum_{k=1}^{n} \Big\{c_{e_j+e_k} x_j x_k + \int e^{-(y,u)}[(x_k y_j + x_j y_k) \sin(x, u)\\ &- x_j x_k \cos(x, u) - y_j y_k \sin^2(\tfrac{1}{2}(x, u))]\nu_{e_j+e_k}(du)\Big\}.\end{aligned}$$

Therefore,

$$\sigma(x, y) = x_j^2 \left[c_{e_j+e_k} + \int e^{-(y,u)} \cos(x, u)\nu_{2e_j}(du)\right] + O(x_j)$$

when $x_j \to +\infty$; but from (6.3.2) and (6.3.3) we have

$$0 \le \sigma(x, y) \le \rho(x, y) = \gamma_j x_j^2 + O(x_j)$$

when $x_j \to +\infty$ where γ_j is the coefficient of z_j^2 in $P(z)$, so that

$$\int e^{-(y,u)} \cos(x, u)\nu_{2e_j}(du) = O(1)$$

when $x_j \to +\infty$ uniformly in x and y. But this integral is the real part of

$$\int e^{i(z,u)}\nu_{2e_j}(du) = \hat{\nu}_{2e_j}(z)$$

and it follows from Corollary A.4.1 that $\hat{\nu}_{2e_j}$ is an entire function of exponential type zero and therefore a constant and, from Theorem 3.6.2, we have

$$\nu_{2e_j} = c'_{2e_j}\delta_0$$

where c'_{2e_j} is a real constant. In the same way, when $x_j = x_k \to +\infty$, we obtain

$$\nu_{e_j+e_k} = c'_{e_j+e_k}\delta_0 .$$

From this and from $\hat{q}(0) = 1$, we obtain (6.3.1) with

$$\beta = (c_{e_1}, \ldots, c_{e_n}), \qquad \nu_j = \nu_{e_j} \qquad (j = 1, \ldots, n),$$

and

$$Q(z) = \sum_{j=1}^{n} \sum_{k=1}^{n} (c_{e_j+e_k} + c'_{e_j+e_k}) z_j z_k .$$

If $Q(x_0) < 0$ for some $x_0 \in R^n$, we easily find that

$$\lim_{r\to\infty} \sigma(rx_0 , y) = -\infty$$

and (6.3.2) is not satisfied. In the same way, if

$$Q(x_0) > P(x_0)$$

for some $x_0 \in R^n$, we have

$$\lim_{r\to\infty} (\rho(rx_0 , y) - \sigma(rx_0 , y)) = -\infty$$

and this contradicts (6.3.2), so that

$$0 \le Q(x) \le P(x)$$

for any $x \in R^n$.

We suppose now that $\tau > 2$. Then

$$\log \hat{q}(iz) = \Phi_\tau(z) + \Psi_\tau(z) + g_\tau(z)$$

where

$$\Phi_\tau(z) = \sum_{k_1+\cdots+k_n=\tau} c_k (iz)^k \qquad \text{and} \qquad \Psi_\tau(z) = \sum_{k_1+\cdots+k_n=\tau} \hat{\nu}_k(z)(iz)^k.$$

Defining ρ and σ as before, we obtain

$$\sigma(x, y) = \sum_{k_1+\cdots+k_n=\tau} \left[c_k + \int e^{-(y,u)} s(x, u)\nu_k(du)\right] x^k + \pi(x, y)$$

where π is a polynomial in x of degree smaller than τ and

$$s(x, u) = \begin{cases} (-1)^{\tau/2} \cos(x, u) & \text{if } \tau \text{ is even,} \\ (-1)^{(\tau-1)/2} \sin(x, u) & \text{if } \tau \text{ is odd.} \end{cases}$$

It follows easily from (6.3.3) that

$$\rho(x, y) = O(\|x\|^2)$$

when $\|x\| \to +\infty$ and from (6.3.2) that

$$\sum_{k_1+\cdots+k_n=\tau} \left[c_k + \int e^{-(y,u)} s(x, u) \nu_k(du)\right] x^k = 0.$$

But this implies as before that

$$\Phi_\tau = \Psi_\tau = 0$$

and the lemma is proved by induction on τ.

Lemma 6.3.2 If, in the preceding lemma, $P(z)$ does not depend on z_j, then $Q(z)$ does not depend on z_j and $\nu_j = 0$. In particular, if, in the preceding lemma, $P = 0$, then

$$\log \hat{q}(z) = i(\beta, z) + \int (e^{i(z,u)} - 1) \nu_0(du)$$

for any $z \in C^n$.

Proof We suppose that $P(z)$ does not depend on z_j and that

$$Q(z) = \sum_{k \le l} d_{k,l} z_k z_l.$$

When $x_j \to +\infty$, using the notations introduced during the proof of the preceding lemma,

$$\begin{aligned} 0 \le \sigma(x, y) = x_j &\left[\sum_{k \le j} d_{k;j} x_k + \sum_{k > j} d_{j,k} x_k \right. \\ &\left. + \int e^{-(y,u)} \sin(x, u) \nu_j(du) \right] + O(1) \\ \le \rho(x, y) = O(1), \end{aligned}$$

so that $Q(z)$ does not depend on z_j and $\nu_j = 0$.

We begin with the induction method and give now the simplest case to which the method can be applied since in this case the idea of the method is not hidden by technical difficulties.

Theorem 6.3.1 Let $\hat{p}$ be an n-variate characteristic function admitting an (α, P, μ)-De Finetti representation where $\mu \in \mathcal{M}_n$ is concentrated on a finite Linnik set Λ constructed on the canonical basis of R^n. If q divides p, then $\hat{q}$ admits the representation

$$\begin{aligned} \hat{q}(t) = i(\beta, t) - Q(t) + \int (e^{i(t,u)} - 1) \nu_0(du) \\ + i \sum_{j=1}^{n} t_j \int (e^{i(t,u)} - 1) \nu_j(du) \end{aligned} \tag{6.3.4}$$

where $\beta \in R^n$, Q is a quadratic form satisfying $0 \leq Q \leq P$, and ν_j are finite signed measures ($j = 0, 1, \ldots, n$). Then ν_0 is concentrated on Λ and ν_j on $\Lambda \cap e_j^\perp$ ($j = 1, \ldots, n$). Moreover, if $P = 0$, then $Q = \nu_j = 0$ ($j = 1, \ldots, n$).

Proof We prove this result by induction on n. The case $n = 1$ follows immediately from Corollary 6.1.1.

If $\Lambda = \Lambda_1 + \cdots + \Lambda_n$, then $\Lambda_j = \{x \in R^n : x = \lambda_{j, m_j} e_j, 0 \leq m_j \leq r_j\}$ is a finite set with $r_j + 1$ elements (we recall that $\lambda_{j,0} = 0$). For any $y = (y_1, \ldots, y_n) \in R^n$, since $\hat{p}$ is an entire characteristic function, $\hat{q}$ is also an entire characteristic function and it follows from Corollaries 3.3.5 and 5.2.2 that the probabilities p_y and q_y defined by

$$\hat{p}_y(t) = \hat{p}(t + iy)/\hat{p}(iy), \qquad \hat{q}_y(t) = \hat{q}(t + iy)/\hat{q}(iy)$$

exist and that q_y divides p_y. From Corollary 2.5.8, we deduce then that $\mathrm{pr}_{e_1} q_y$ divides $\mathrm{pr}_{e_1} p_y$ and $\mathrm{pr}_{e_1^\perp} q_y$ divides $\mathrm{pr}_{e_1^\perp} p_y$ for any $y \in R^n$.

Since $\mathrm{pr}_{e_1} p_y$ satisfies the conditions of Corollary 6.1.1, we deduce from this result that $\mathrm{pr}_{e_1} q_y$ admits a Lévy–Hinčin representation with a Poisson measure concentrated on Λ_1 and this is equivalent to the relation

$$\log \hat{q}(z_1, iy_2, \ldots, iy_n) = i\alpha(y_2, \ldots, y_n)z_1 - \beta(y_2, \ldots, y_n)z_1^2 + \sum_{m=0}^{r_1} \gamma_m(y_2, \ldots, y_n) \exp(i\lambda_{1,m} z_1) \tag{6.3.5}$$

valid for $z_1 \in C$, $(y_2, \ldots, y_n) \in R^{n-1}$ where α, β, and γ_m are functions defined on R^{n-1} with real values.

Since $\mathrm{pr}_{e_1^\perp} p_y$ is (up to an isomorphism) an $(n-1)$-variate probability satisfying the conditions of Theorem 6.3.1, it follows by induction hypothesis that

$$\begin{aligned}\log \hat{q}(iy_1, z_2, \ldots, z_n) = {} & \sum_{2 \leq j \leq k \leq n} a_{j,k}(y_1) z_j z_k \\ & + \sum_{m_2=0}^{r_2} \cdots \sum_{m_n=0}^{r_n} b_{m_2, \ldots, m_n}(y_1) \exp\left(i \sum_{j=2}^{n} \lambda_{j, m_j} z_j\right) \\ & + i \sum_{k=2}^{n} z_k \left(\sum_{m_2=0}^{r_2} \cdots \sum_{m_{k-1}=0}^{r_{k-1}} \sum_{m_{k+1}=0}^{r_{k+1}} \cdots \sum_{m_n=0}^{r_n} \right. \\ & \left. \times c_{k, m_2, \ldots, \widehat{m_k}, \ldots, m_n}(y_1) \exp\left(i \sum_{\substack{2 \leq j \leq n \\ j \neq k}} \lambda_{j, m_j} z_j\right)\right)\end{aligned} \tag{6.3.6}$$

holds for $y_1 \in R$, $(z_2, \ldots, z_n) \in C^{n-1}$ where the a, b, and c are functions with real values (the caret means that the corresponding term is missing). Comparing (6.3.5) and (6.3.6), we obtain the equation

$$-\alpha(y_2, \ldots, y_n)y_1 + \beta(y_2, \ldots, y_n)y_1^2 + \sum_{m=0}^{r_1} \gamma_m(y_2, \ldots, y_n)\exp(-\lambda_{1,m}y_1)$$
$$= -\sum_{2\le j\le k\le n} a_{j,k}(y_1)y_jy_k + \sum_{m_2=0}^{r_2}\cdots\sum_{m_n=0}^{r_n} b_{m_2,\ldots,m_n}(y_1) \times \exp\left(-\sum_{j=2}^{n}\lambda_{j,m_j}y_j\right) - \sum_{k=2}^{n} y_k \sum_{m_2=0}^{r_2}\cdots\sum_{m_{k-1}=0}^{r_{k-1}}$$
$$\times \sum_{m_{k+1}=0}^{r_{k+1}}\cdots\sum_{m_n=0}^{r_n} c_{k,m_2,\ldots,\widehat{m_k},\ldots,m_n}(y_1)$$
$$\times \exp(-(\lambda_{2,m_2}y_2 + \cdots + \widehat{\lambda_{k,m_k}y_k} + \cdots + \lambda_{n,m_n}y_n)) \quad (6.3.7)$$

for any $(y_1, \ldots, y_n) \in R^n$. Since y_1, y_1^2, and $\exp(-\lambda_{1,m}y_1)$ are $r_1 + 3$ linearly independent functions, we can find $r_1 + 3$ values of y_1 such that the system of equations obtained by replacing y_1 by these values in (6.3.7) is a Cramer system of $r_1 + 3$ linear equations with $r_1 + 3$ unknown: α, β, and γ_m $(m = 0, 1, \ldots, r_1)$. Solving this system, we obtain

$$-\alpha(y_2, \ldots, y_n) = \sum_{2\le j\le k\le n} a_{j,k}^{(\alpha)}y_jy_k + \sum_{m_2=0}^{r_2}\cdots\sum_{m_n=0}^{r_n} b_{m_2,\ldots,m_n}^{(\alpha)} \times \exp\left(-\sum_{j=2}^{n}\lambda_{j,m_j}y_j\right)$$
$$+ \sum_{k=2}^{n} y_k\left[\sum_{m_2=0}^{r_2}\cdots\sum_{m_{k-1}=0}^{r_{k-1}}\sum_{m_{k+1}=0}^{r_{k+1}}\cdots\sum_{m_n=0}^{r_n} c_{k,m_2,\ldots,\widehat{m_k},\ldots,m_n}^{(\alpha)} \times \exp(-(\lambda_{2,m_2}y_2 + \cdots + \widehat{\lambda_{k,m_k}y_k} + \cdots + \lambda_{n,m_n}y_n))\right],$$

$$\beta(y_2, \ldots, y_n) = \sum_{2\le j\le k\le n} a_{j,k}^{(\beta)}y_jy_k + \sum_{m_2=0}^{r_2}\cdots\sum_{m_n=0}^{r_n} b_{m_2,\ldots,m_n}^{(\beta)} \times \exp\left(-\sum_{j=2}^{n}\lambda_{j,m_j}y_j\right)$$
$$+ \sum_{k=2}^{n} y_k\left[\sum_{m_2=0}^{r_2}\cdots\sum_{m_{k-1}=0}^{r_{k-1}}\sum_{m_{k+1}=0}^{r_{k+1}}\cdots\sum_{m_n=0}^{r_n} c_{k,m_2,\ldots,\widehat{m_k},\ldots,m_n}^{(\beta)} \times \exp(-(\lambda_{2,m_2}y_2 + \cdots + \widehat{\lambda_{k,m_k}y_k} + \cdots + \lambda_{n,m_n}y_n))\right],$$

$$\gamma_m(y_2, \ldots, y_n) = \sum_{2 \le j \le k \le n} a_{j,k}^{(\gamma_m)} y_j y_k + \sum_{m_2=0}^{r_2} \cdots \sum_{m_n=0}^{r_n} b_{m_2, \ldots, m_n}^{(\gamma_m)}$$
$$\times \exp\left(-\sum_{j=2}^{n} \lambda_{j, m_j} y_j\right)$$
$$+ \sum_{k=2}^{n} y_k \left[\sum_{m_2=0}^{r_2} \cdots \sum_{m_{k-1}=0}^{r_{k-1}} \sum_{m_{k+1}=0}^{r_{k+1}} \cdots \sum_{m_n=0}^{r_n} c_{k, m_2, \ldots, \widehat{m_k}, \ldots, m_n}^{(\gamma_m)} \right.$$
$$\left. \times \exp(-(\lambda_{2, m_2} y_2 + \cdots + \widehat{\lambda_{k, m_k} y_k} + \cdots + \lambda_{n, m_n} y_n)) \right]$$

where the a, b, and c are real constants. Substituting these values in (6.3.4), we obtain

$$\log \hat{q}(iy) = -(\beta, y) + Q(y) + \Pi_3(y) + \Pi_4(y)$$
$$+ \int (e^{-(y,u)} - 1)\nu_0(du) - \sum_{j=1}^{n} y_j \int (e^{-(y,u)} - 1)\nu_j(du)$$
$$+ \sum_{1 \le j \le k \le n} y_j y_k \int (e^{-(y,u)} - 1)\nu'_{j,k}(du)$$
$$+ \sum_{k=2}^{n} y_1^2 y_k \int (e^{-(y,u)} - 1)\nu''_k(du) \tag{6.3.8}$$

for any $y \in R^n$ where

$$\beta = -(b_{0, \ldots, 0}^{(\alpha)}, c_{2, 0, \ldots, 0}^{(\gamma_0)}, \ldots, c_{n, 0, \ldots, 0}^{(\gamma_0)}),$$

$$Q(y) = \sum_{2 \le j \le k \le n} a_{j,k}^{(\gamma_0)} y_j y_k + \sum_{k=2}^{n} c_{k, 0, \ldots, 0}^{(\alpha)} y_1 y_k + b_{0, \ldots, 0}^{(\beta)} y_1^2,$$

$$\Pi_3(y) = \sum_{2 \le j \le k \le n} a_{j,k}^{(\alpha)} y_1 y_j y_k + \sum_{k=2}^{n} c_{k, 0, \ldots, 0}^{(\beta)} y_1^2 y_k,$$

$$\Pi_4(y) = \sum_{2 \le j \le k \le n} a_{j,k}^{(\beta)} y_1^2 y_j y_k,$$

$$\nu_0(B) = \sum_{(\lambda_{1,m_1}, \ldots, \lambda_{n,m_n}) \in B} b_{m_2, \ldots, m_n}^{(\gamma_{m_1})},$$

$$\nu_1(B) = -\sum_{(0, \lambda_{2,m_2}, \ldots, \lambda_{n,m_n}) \in B} b_{m_2, \ldots, m_n}^{(\alpha)},$$

$$\nu_k(B) = -\sum_{\substack{(\lambda_{1,m_1}, \ldots, \lambda_{k-1,m_{k-1}}, 0, \\ \lambda_{k+1,m_{k+1}}, \ldots, \lambda_{n,m_n}) \in B}} c_{k, m_2, \ldots, \widehat{m_k}, \ldots, m_n}^{(\gamma_{m_1})} \qquad (k = 2, \ldots, n),$$

$$\nu'_{1,1}(B) = \sum_{(0, \lambda_{2,m_2}, \ldots, \lambda_{n,m_n}) \in B} b_{m_2, \ldots, m_n}^{(\beta)},$$

$$v'_{1,k}(B) = \sum_{\substack{(0,\, \lambda_{2,m_2},\, \ldots,\, \lambda_{k-1,m_{k-1}},\, 0, \\ \lambda_{k+1,m_{k+1}},\, \ldots,\, \lambda_{n,m_n}) \in B}} c^{(\alpha)}_{k, m_2, \ldots, \widehat{m_k}, \ldots, m_n} \qquad (k = 2, \ldots, n),$$

$$v'_{j,k}(B) = \sum_{(\lambda_{1,m},\, 0,\, \ldots,\, 0) \in B} a^{(\gamma_m)}_{j,k} \qquad (2 \le j \le k \le n),$$

$$v''_k(B) = \sum_{\substack{(0,\, \lambda_{2,m_2},\, \ldots,\, \lambda_{k-1,m_{k-1}},\, 0, \\ \lambda_{k+1,m_{k+1}},\, \ldots,\, \lambda_{n,m_n}) \in B}} c^{(\beta)}_{k, m_2, \ldots, \widehat{m_k}, \ldots, m_n} \qquad (k = 2, \ldots, n)$$

for any $B \in \mathscr{B}_n$. Since the right side of (6.3.8) is an entire function, we find that (6.3.8) holds for any $y \in C^n$. Now we can apply Lemma 6.3.1 and find

$$\Pi_3 = \Pi_4 = v'_{j,k} = v''_k = 0, \qquad 0 \le Q \le P.$$

If $P = 0$, we deduce from Lemma 6.3.2 that

$$v_j = 0 \qquad (j = 1, \ldots, n)$$

and this proves the theorem.

We extend now the preceding result to the case of an infinite Poisson spectrum and begin with

Theorem 6.3.2 Let $\hat{p}$ be an infinitely divisible characteristic function admitting an (α, P, μ)-De Finetti representation with a measure μ satisfying the following conditions:

(a) μ is concentrated on a Linnik set $\Lambda = \Lambda_1 + \cdots + \Lambda_n$ constructed on the canonical basis of R^n, $\Lambda_j = \{\lambda_{k,j} e_j\}$ having no finite accumulation point $(j = 1, \ldots, n)$;

(b) for some $K > 0$,

$$\mu(\{x : \|x\| > u\}) = O(\exp(-Ku^2)) \qquad (u \to +\infty).$$

If q divides p, then $\hat{q}$ admits representation (6.3.4) for any $t \in C^n$ where $\beta \in R^n$, Q is a quadratic form satisfying $0 \le Q \le P$ and $v_j \in \mathscr{M}_n$ are such that

$$\int e^{i(t,u)} v_j(du)$$

are entire functions of t $(j = 0, 1, \ldots, n)$. Then v_0 is concentrated on Λ and v_j on $\Lambda \cap e_j^\perp$ $(j = 1, \ldots, n)$. Moreover, if $P = 0$, then $v_j = 0$ $(j = 1, \ldots, n)$.

Proof The method is very similar to the one used in Theorem 6.3.1. We indicate only the differences. We deduce from Theorem 3.3.4 that $\hat{p}$ is an entire characteristic function (more precisely, $\log \hat{p}$ is an entire function of order 2) and obtain from this, exactly as in the preceding theorem, the representations

$$\log \hat{q}(z_1, iy_2, \ldots, iy_n) = i\alpha(y_2, \ldots, y_n)z_1 - \beta(y_2, \ldots, y_n)y_1^2 + \sum_{m=0}^{\infty} \gamma_m(y_2, \ldots, y_n) \exp(i\lambda_{1,m} z_1)$$

valid for $z_1 \in C$, $(y_2, \ldots, y_n) \in R^{n-1}$ where α, β, and γ_m are functions defined on R^{n-1} with real values and

$$\begin{aligned} \log \hat{q}(iy_1, z_2, \ldots, z_n) = {} & \sum_{2 \le j \le k \le n} a_{j,k}(y_1) z_j z_k \\ & + \sum_{m_2=0}^{\infty} \cdots \sum_{m_n=0}^{\infty} b_{m_2, \ldots, m_n}(y_1) \exp\left(i \sum_{j=2}^{n} \lambda_{j, m_j} z_j\right) \\ & + i \sum_{k=2}^{n} z_k \left[\sum_{m_2=0}^{\infty} \cdots \sum_{m_{k-1}=0}^{\infty} \sum_{m_{k+1}=0}^{\infty} \cdots \right. \\ & \left. \times \sum_{m_n=0}^{\infty} c_{k, m_2, \ldots, \widehat{m_k}, \ldots, m_n}(y_1) \exp\left(i \sum_{\substack{2 \le j \le n \\ j \ne k}} \lambda_{j, m_j} z_j\right) \right] \end{aligned}$$

valid for $y_1 \in R$, $(z_2, \ldots, z_n) \in C^{n-1}$ where the a, b, and c are functions with real values. From this, we obtain the equation

$$\begin{aligned} & -\alpha(y_2, \ldots, y_n)y_1 + \beta(y_2, \ldots, y_n)y_1^2 + \sum_{m=0}^{\infty} \gamma_m(y_2, \ldots, y_n) \\ & \times \exp(-\lambda_{1,m} y_1) = - \sum_{2 \le j \le k \le n} a_{j,k}(y_1) y_j y_k \\ & + \sum_{m_2=0}^{\infty} \cdots \sum_{m_n=0}^{\infty} b_{m_2, \ldots, m_n}(y_1) \exp\left(-\sum_{j=2}^{n} \lambda_{j, m_j} y_j\right) \\ & - \sum_{k=2}^{n} y_k \left[\sum_{m_2=0}^{\infty} \cdots \sum_{m_{k-1}=0}^{\infty} \sum_{m_{k+1}=0}^{\infty} \cdots \sum_{m_n=0}^{\infty} \right. \\ & \left. \times c_{k, m_2, \ldots, \widehat{m_k}, \ldots, m_n}(y_1) \exp\left(-\sum_{\substack{2 \le j \le n \\ j \ne k}} \lambda_{j, m_j} y_j\right) \right] \end{aligned}$$

valid for any $y = (y_1, \ldots, y_n) \in R^n$.

If $\varepsilon > 0$, $v = (v_1, \ldots, v_n) \in R^n$, $v' = (v'_1, \ldots, v'_n) \in R^n$ $(v < v')$ are given, we can find $r = (r_1, \ldots, r_n) \in N^n$ such that the preceding equation becomes

$$
\begin{aligned}
&-\alpha(y_2, \ldots, y_n)y_1 + \beta(y_2, \ldots, y_n)y_1^2 + \sum_{m=0}^{r_1} \gamma_m(y_2, \ldots, y_n) \\
&\quad \times \exp(-\lambda_{1,m}y_1) = -\sum_{2 \le j \le k \le n} a_{j,k}(y_1)y_j y_k \\
&\quad + \sum_{m_2=0}^{r_2} \cdots \sum_{m_n=0}^{r_n} b_{m_2, \ldots, m_n}(y_1) \exp\left(-\sum_{j=2}^{n} \lambda_{j,m_j} y_j\right) \\
&\quad - \sum_{k=2}^{n} y_k \left[\sum_{m_2=0}^{r_2} \cdots \sum_{m_{k-1}=0}^{r_{k-1}} \sum_{m_{k+1}=0}^{r_{k+1}} \cdots \sum_{m_n=0}^{r_n} \right. \\
&\quad \left. \times\, c_{k, m_2, \ldots, \widehat{m_k}, \ldots, m_n}(y_1) \exp\left(-\sum_{\substack{2 \le j \le n \\ j \ne k}} \lambda_{j,m_j} y_j\right) \right] + R_\varepsilon(y)
\end{aligned}
$$

where $|R_\varepsilon(y)| < \varepsilon$ for any $v \le y \le v'$. We can solve this equation exactly as in the preceding theorem and obtain

$$
\begin{aligned}
\alpha(y_2, \ldots, y_n) &= \sum_{2 \le j \le k \le n} a^{(\alpha)}_{\varepsilon, j, k} y_j y_k + \sum_{m_2=0}^{r_2} \cdots \sum_{m_n=0}^{r_n} b^{(\alpha)}_{\varepsilon, m_2, \ldots, m_n} \\
&\quad \times \exp\left(-\sum_{j=2}^{n} \lambda_{j,m_j} y_j\right) + \sum_{k=2}^{n} y_k \left[\sum_{m_2=0}^{r_2} \cdots \sum_{m_{k-1}=0}^{r_{k-1}} \sum_{m_{k+1}=0}^{r_{k+1}} \cdots \right. \\
&\quad \left. \times \sum_{m_n=0}^{r_n} c^{(\alpha)}_{\varepsilon, k, m_2, \ldots, \widehat{m_k}, \ldots, m_n} \exp\left(-\sum_{\substack{2 \le j \le n \\ j \ne k}} \lambda_{j,m_j} y_j\right) \right] \\
&\quad + R^{(\alpha)}_\varepsilon(y_2, \ldots, y_n), \\
\beta(y_2, \ldots, y_n) &= \sum_{2 \le j \le k \le n} a^{(\beta)}_{\varepsilon, j, k} y_j y_k + \sum_{m_2=0}^{r_2} \cdots \sum_{m_n=0}^{r_n} b^{(\beta)}_{\varepsilon, m_2, \ldots, m_n} \\
&\quad \times \exp\left(-\sum_{j=2}^{n} \lambda_{j,m_j} y_j\right) + \sum_{k=2}^{n} y_k \left[\sum_{m_2=0}^{r_2} \cdots \sum_{m_{k-1}=0}^{r_{k-1}} \sum_{m_{k+1}=0}^{r_{k+1}} \cdots \right. \\
&\quad \left. \times \sum_{m_n=0}^{r_n} c^{(\beta)}_{\varepsilon, k, m_2, \ldots, \widehat{m_k}, \ldots, m_n} \exp\left(-\sum_{\substack{2 \le j \le n \\ j \ne k}} \lambda_{j,m_j} y_j\right) \right] \\
&\quad + R^{(\beta)}_\varepsilon(y_2, \ldots, y_n),
\end{aligned}
$$

$$\gamma_m(y_2, \dots, y_n) = \sum_{2 \le j \le k \le n} a^{(\gamma_m)}_{\varepsilon, j, k} y_j y_k + \sum_{m_2=0}^{r_2} \cdots \sum_{m_n=0}^{r_n} b^{(\gamma_m)}_{\varepsilon, m_2, \dots, m_n}$$
$$\times \exp\left(-\sum_{j=2}^{n} \lambda_{j, m_j} y_j\right) + \sum_{k=2}^{n} y_k \left[\sum_{m_2=0}^{r_2} \cdots \sum_{m_{k-1}=0}^{r_{k-1}} \sum_{m_{k+1}=0}^{r_{k+1}} \cdots\right.$$
$$\left.\times \sum_{m_n=0}^{r_n} c^{(\gamma_m)}_{\varepsilon, k, m_2, \dots, \widehat{m_k}, \dots, m_n} \exp\left(-\sum_{\substack{2 \le j \le n \\ j \ne k}} \lambda_{j, m_j} y_j\right)\right]$$
$$+ R^{(\gamma_m)}_\varepsilon(y_2, \dots, y_n)$$

where

$$R^{(\alpha)}_\varepsilon(y_2, \dots, y_n) y_1 + R^{(\beta)}_\varepsilon(y_2, \dots, y_n) y_1^2$$
$$+ \sum_{m=0}^{r_1} R^{(\gamma_m)}_\varepsilon(y_2, \dots, y_n) \exp(-\lambda_{1, m} y_1) = R_\varepsilon(y_1, \dots, y_n)$$

for $v \le y = (y_1, \dots, y_n) \le v'$.

Since Λ_j has no finite accumulation point, the set $\{y_j, y_j^2, \exp(-\lambda_{j, m_j} y_j),$ $m = 0, 1, 2, \dots\}$ is a set of topologically independent functions on $[v_j, v_j']$ (see Appendix A.5) ($j = 1, \dots, n$). Letting $\varepsilon \to 0$ and using this fact for $j = 1$, we obtain

$$\lim_{\varepsilon \to 0} R^{(\alpha)}_\varepsilon(y_2, \dots, y_n) = \lim_{\varepsilon \to 0} R^{(\beta)}_\varepsilon(y_2, \dots, y_n) = \lim_{\varepsilon \to 0} R^{(\gamma_m)}_\varepsilon(y_2, \dots, y_n) = 0,$$

and this uniformly with respect to $(y_2, \dots, y_n) \in \prod_{j=2}^n [v_j, v_j']$. From the topological independence of

$$\left\{y_j y_k, \exp\left(-\sum_{j=2}^{n} \lambda_{j, m_j} y_j\right), y_k \exp(-(\lambda_{2, m_2} y_2 + \cdots + \widehat{\lambda_{k, m_k} y_k} + \cdots + \lambda_{n, m_n} y_n))\right\}$$

on $\prod_{j=2}^n [v_j, v_j']$, we obtain then the representation

$$\alpha(y_2, \dots, y_n) = \sum_{2 \le j \le k \le n} a^{(\alpha)}_{j, k} y_j y_k$$
$$+ \sum_{m_2=0}^{\infty} \cdots \sum_{m_n=0}^{\infty} b^{(\alpha)}_{m_2, \dots, m_n} \exp\left(-\sum_{j=2}^{n} \lambda_{j, m_j} y_j\right)$$
$$+ \sum_{k=2}^{n} y_k \left[\sum_{m_2=0}^{\infty} \cdots \sum_{m_{k-1}=0}^{\infty} \sum_{m_{k+1}=0}^{\infty} \cdots\right.$$
$$\left.\times \sum_{m_n=0}^{\infty} c^{(\alpha)}_{k, m_2, \dots, \widehat{m_k}, \dots, m_n} \exp\left(-\sum_{\substack{2 \le j \le n \\ j \ne k}} \lambda_{j, m_j} y_j\right)\right] \tag{6.3.9}$$

and similar representations for β and γ_m where the a, b, and c are real constants. Using again the topological independence of $\{y_j y_k, \dots\}$, we see that this representation does not depend on v and v' and therefore holds for any $(y_2, \dots, y_n) \in R^{n-1}$. Therefore the series in (6.3.9) converges for any $(y_2, \dots, y_n) \in C^{n-1}$ and (6.3.9) defines α as an entire function of $(y_2, \dots, y_n)$. Since we can do the same for β and γ_m, we obtain finally for $\log \hat{q}(iy)$ representation (6.3.8) for any $y \in C^n$ where β, Q, Π_3, Π_4, v_j, $v'_{j,k}$, and v''_k are defined as in the preceding theorem. The result follows then from Lemmas 6.3.1 and 6.3.2.

Theorem 6.3.3 Let $\hat{p}$ be an infinitely divisible characteristic function admitting an (α, P, μ)-Lévy–Hinčin representation with a measure μ satisfying the following conditions:

(a) μ is concentrated on a Linnik set $\Lambda = \Lambda_1 + \cdots + \Lambda_n$ constructed on the canonical basis of R^n, $\Lambda_j = \{\lambda_{j, m_j} e_j\}$ being a finite set with $r_j + 1$ elements $(j = 2, \dots, n)$;

(b) for some $K > 0$,

$$\mu(\{x : \|x\| > u\}) = O(\exp(-Ku^2)) \qquad (u \to +\infty).$$

If q divides p, then $\hat{q}$ admits for any $t \in C^n$ the representation

$$\log \hat{q}(t) = i(\beta, t) - Q(t) + \int K(t, u) v_0(du) + i \sum_{j=1}^{n} \int K(t, u) v_j(du) t_j$$

where $\beta \in R^n$, Q is a quadratic form, and v_j are finite signed measures satisfying

$$v_j(\{x : \|x\| > u\}) = O(\exp(-Ku^2)) \qquad (u \to +\infty)$$

$(j = 0, 1, \dots, n)$. Then v_0 is concentrated on Λ and v_j on $\Lambda \cap e_j^{\perp}$ $(j = 1, \dots, n)$. Moreover, if $P = 0$, then $Q = v_j = 0$ $(j = 1, \dots, n)$.

Proof From Theorems 3.6.5 and 3.3.4, we deduce that $\hat{p}$ is an entire characteristic function and from this and Theorem 6.3.1, we have the representation

$$\begin{aligned}
\log \hat{q}(iy_1, z_2, \dots, z_n) = & \sum_{2 \le j \le k \le n} a_{j,k}(y_1) z_j z_k \\
& + \sum_{m_2=0}^{r_2} \cdots \sum_{m_n=0}^{r_n} b_{m_2, \dots, m_n}(y_1) \exp\left(i \sum_{j=2}^{n} \lambda_{j, m_j} y_j \right) \\
& + i \sum_{k=2}^{n} z_k \left[\sum_{m_2=0}^{r_2} \cdots \sum_{m_{k-1}=0}^{r_{k-1}} \sum_{m_{k+1}=0}^{r_{k+1}} \cdots \right. \\
& \left. \times \sum_{m_n=0}^{r_n} c_{k, m_2, \dots, \widehat{m_k}, \dots, m_n}(y_1) \exp\left(i \sum_{\substack{2 \le j \le n \\ j \ne k}} \lambda_{j, m_j} z_j \right) \right]
\end{aligned} \qquad (6.3.10)$$

for any $y_1 \in R$, $(z_2, \dots, z_n) \in C^{n-1}$ where the a, b, and c are functions with real values. From Theorem 6.1.1, we obtain the representation

$$\log \hat{q}(z_1, iy_2, \ldots, iy_n) = \alpha(y_2, \ldots, y_n) + i\beta(y_2, \ldots, y_n)z_1$$
$$+ \gamma(y_2, \ldots, y_n)z_1^2 + \sum_{m=1}^{\infty} \delta_m(y_2, \ldots, y_n)K_1(\lambda_{1,m}, z_1) \quad (6.3.11)$$

for $z_1 \in C$, $(y_2, \ldots, y_n) \in R^{n-1}$ where

$$K_1(\lambda, x) = e^{i\lambda x} - 1 - i\lambda x(1 + \lambda^2)^{-1}$$

and the α, β, γ, and δ_m are functions defined on R^{n-1} with real values and satisfying

$$\sum_{m=1}^{\infty} |\delta_m(y_2, \ldots, y_n)| (1 + \lambda_{1,m}^2)\lambda_{1,m}^{-2} < +\infty,$$
$$\sum_{|\lambda_{1,m}| > u} |\delta_m(y_2, \ldots, y_n)| = O(\exp(-Ku^2)) \qquad (u \to +\infty)$$

uniformly for $(y_2, \ldots, y_n)$ belonging to a compact set of R^{n-1}. From (6.3.10) and (6.3.11), we obtain the equation

$$-\sum_{2 \le j \le k \le n} a_{j,k} y_j y_k + \sum_{m_2=0}^{r_2} \cdots \sum_{m_n=0}^{r_n} b_{m_2, \ldots, m_n}(y_1)$$
$$\times \exp\left(-\sum_{j=2}^{n} \lambda_{j, m_j} y_j\right) - \sum_{k=2}^{n} y_k \left[\sum_{m_2=0}^{r_2} \cdots \sum_{m_{k-1}=0}^{r_{k-1}} \sum_{m_{k+1}=0}^{r_{k+1}} \cdots\right.$$
$$\left.\times \sum_{m_n=0}^{r_n} c_{k, m_2, \ldots, \widehat{m_k}, \ldots, m_n}(y_1) \exp\left(-\sum_{\substack{2 \le j \le n \\ j \ne k}} \lambda_{j, m_j} y_j\right)\right]$$
$$= \alpha(y_2, \ldots, y_n) - \beta(y_2, \ldots, y_n)y_1 - \gamma(y_2, \ldots, y_n)y_1^2$$
$$+ \sum_{m=1}^{\infty} \delta_m(y_2, \ldots, y_n)K_1(\lambda_{1,m}, iy_1)$$

for any $(y_1, \ldots, y_n) \in R^n$ and using the linear independence of

$$\left\{y_j y_k, \exp\left(-\sum_{j=2}^{n} \lambda_{j, m_j} y_j\right), y_k \exp\left(-\sum_{\substack{2 \le j \le n \\ j \ne k}} \lambda_{j, m_j} y_j\right)\right\},$$

we can solve this equation and obtain

$$a_{j,k} = \alpha_{j,k}^{(a)} + \beta_{j,k}^{(a)} y_1 + \gamma_{j,k}^{(a)} y_1^2 + \sum_{m=1}^{\infty} \delta_{m,j,k}^{(a)} K_1(\lambda_{1,m}, iy_1),$$
$$b_{m_2, \ldots, m_n} = \alpha_{m_2, \ldots, m_n}^{(b)} + \beta_{m_2, \ldots, m_n}^{(b)} y_1 + \gamma_{m_2, \ldots, m_n}^{(b)} y_1^2$$
$$+ \sum_{m=1}^{\infty} \delta_{m, m_2, \ldots, m_n}^{(b)} K_1(-\lambda_{1,m}, iy_1),$$

$$c_{k, m_2, \ldots, \widehat{m_k}, \ldots, m_n} = \alpha^{(c_k)}_{m_2, \ldots, \widehat{m_k}, \ldots, m_n} + \beta^{(c_k)}_{m_2, \ldots, \widehat{m_k}, \ldots, m_n} y_1 + \gamma^{(c_k)}_{m_2, \ldots, \widehat{m_k}, \ldots, m_n} y_1^2 + \sum_{m=1}^{\infty} \delta^{(c_k)}_{m_2, \ldots, \widehat{m_k}, \ldots, m_n} K_1(-\lambda_{1, m}, iy_1)$$

where the α, β, γ, and δ are real constants satisfying

$$\sum_{m=1}^{\infty} \delta^{(a)}_{m, j, k}(1 + \lambda^2_{1, m})\lambda^{-2}_{1, m} < +\infty,$$

$$\sum_{m=1}^{\infty} \delta^{(b)}_{m, m_2, \ldots, m_n}(1 + \lambda^2_{1, m})\lambda^{-2}_{1, m} < +\infty, \tag{6.3.12}$$

$$\sum_{m=1}^{\infty} \delta^{(c_k)}_{m, m_2, \ldots, m_n}(1 + \lambda^2_{1, m})\lambda^{-2}_{1, m} < +\infty,$$

$$\sum_{|\lambda_{1,m}| > u} |\delta^{(a)}_{m, j, k}| = O(\exp(-Ku^2))$$

$$\sum_{|\lambda_{1,m}| > u} |\delta^{(b)}_{m, m_2, \ldots, m_n}| = O(\exp(-Ku^2)) \tag{6.3.13}$$

$$\sum_{|\lambda_{1,m}| > u} |\delta^{(c_k)}_{m, m_2, \ldots, m_k, \ldots, m_n}| = O(\exp(-Ku^2)).$$

Substituting these values in (6.3.10), we obtain

$$\begin{aligned}
\log \hat{q}(iy) = {} & \sum_{2 \le j \le k \le n} \Bigg\{ \alpha^{(a)}_{j, k} + \beta^{(a)}_{j, k} y_1 + \gamma^{(a)}_{j, k} y_1^2 \\
& + \sum_{m=1}^{\infty} \delta^{(a)}_{m, j, k} K_1(-\lambda_{1, m}, iy_1) \Bigg\} y_j y_k \\
& + \sum_{m_2=0}^{r_2} \cdots \sum_{m_n=0}^{r_n} \Bigg\{ \alpha^{(b)}_{m_2, \ldots, m_n} + \beta^{(b)}_{m_2, \ldots, m_n} y_1 + \gamma^{(b)}_{m_2, \ldots, m_n} y_1^2 \\
& + \sum_{m=1}^{\infty} \delta_{m, m_2, \ldots, m_n} K_1(-\lambda_{1, m}, iy_1) \Bigg\} \exp\left(-\sum_{j=2}^{n} \lambda_{j, m} y_j\right) \\
& + \sum_{k=2}^{n} y_k \Bigg[\sum_{m_2=0}^{r_2} \cdots \sum_{m_{k-1}=0}^{r_{k-1}} \sum_{m_{k+1}=0}^{r_{k+1}} \cdots \sum_{m_n=0}^{r_n} \Bigg\{ \alpha^{(c_k)}_{m_2, \ldots, \widehat{m_k}, \ldots, m_n} \\
& + \beta^{(c_k)}_{m_2, \ldots, \widehat{m_k}, \ldots, m_n} y_1 + \gamma^{(c_k)}_{m_2, \ldots, \widehat{m_k}, \ldots, m_n} y_1^2 \\
& + \sum_{m=1}^{\infty} \delta^{(c_k)}_{m, m_2, \ldots, \widehat{m_k}, \ldots, m_n} K_1(-\lambda_{1, m}, iy_1) \Bigg\} \exp\Bigg(-\sum_{\substack{2 \le j \le n \\ j \ne k}} \lambda_{j, m_j} y_j\Bigg) \Bigg]
\end{aligned} \tag{6.3.14}$$

for any $y \in R^n$. From (6.3.12) and (6.3.13), we deduce that the right side of this equation is an entire function of y, so that (6.3.14) holds for any $y \in C^n$.

Defining ρ and σ by

$$\rho(x, y) = \operatorname{Re}(\log \hat{p}(iy) - \log \hat{p}(x + iy)),$$
$$\sigma(x, y) = \operatorname{Re}(\log \hat{q}(iy) - \log \hat{q}(x + iy))$$

for any $x \in R^n$, $y \in R^n$, we have, from Corollary 5.2.3,

$$0 \leq \sigma(x, y) \leq \rho(x, y) \tag{6.3.15}$$

for any $x \in R^n$, $y \in R^n$ and from the definition

$$\rho(x, y) = P(x) + \int e^{-(y, u)} \sin^2(\tfrac{1}{2}(x, u)) \frac{1 + \|u\|^2}{\|u\|^2} \mu(du).$$

From these relations, we deduce exactly as in Lemma 6.3.1 that

$$\gamma_{j,k}^{(a)} = 0,$$
$$\delta_{m,j,k}^{(a)} = 0 \quad \text{if} \quad m \neq 0,$$
$$\gamma_{m_2,\ldots,m_n}^{(b)} = 0 \quad \text{if} \quad (m_2, \ldots, m_n) \neq 0,$$
$$\beta_{m_2,\ldots,\widehat{m_k},\ldots,m_n}^{(c_k)} = \gamma_{m_2,\ldots,\widehat{m_k},\ldots,m_n}^{(c_k)} = 0 \quad \text{if} \quad (m_2, \ldots, \widehat{m_k}, \ldots, m_n) \neq 0.$$

Using the identity

$$\begin{aligned}
&[\exp(\lambda_{1,m_1} z_1) - 1 - \lambda_{1,m_1} z_1 (1 + \lambda_{1,m_1}^2)^{-1}] \exp\left(\sum_{j=2}^{n} \lambda_{j,m_j} z_j\right) \\
&= \exp\left(\sum_{j=1}^{n} \lambda_{j,m_j} z_j\right) - 1 - \left(\sum_{j=1}^{n} \lambda_{j,m_j} z_j\right)\left(1 + \sum_{j=1}^{n} \lambda_{j,m_j}^2\right)^{-1} \\
&\quad - \left[\exp\left(\sum_{j=2}^{n} \lambda_{j,m_j} z_j\right) - 1 - \left(\sum_{j=2}^{n} \lambda_{j,m_j} z_j\right)\left(1 + \sum_{j=2}^{n} \lambda_{j,m_j}^2\right)^{-1}\right] \\
&\quad \times [\lambda_{1,m_1} z_1 (1 + \lambda_{1,m_1}^2)^{-1} + 1] - \lambda_{1,m_1}^2 \left(\sum_{j=2}^{n} \lambda_{j,m_j} z_j\right) \\
&\quad \times \left(1 + \sum_{j=2}^{n} \lambda_{j,m_j}^2\right)^{-1} \left(1 + \sum_{j=1}^{n} \lambda_{j,m_j}^2\right)^{-1} - \lambda_{1,m_1} z_1 (1 + \lambda_{1,m_1}^2)^{-1} \\
&\quad \times \left[\left(\sum_{j=2}^{n} \lambda_{j,m_j}^2\right)\left(1 + \sum_{j=1}^{n} \lambda_{j,m_j}^2\right)^{-1} + \left(\sum_{j=2}^{n} \lambda_{j,m_j} z_j\right)\left(1 + \sum_{j=2}^{n} \lambda_{j,m_j}^2\right)^{-1}\right]
\end{aligned}$$

we obtain from (6.3.14) and from $\log \hat{q}(0) = 0$

$$\log \hat{q}(iy) = \sum_{j=1}^{3} \Pi_j(y) + \int K(iy, u) \nu_0(du) - \sum_{j=1}^{n} y_j \int K(iy, u) \nu_j(du)$$

for any $y \in C^n$ where Π_j is a polynomial with degree j and with real coefficients and ν_j are finite signed measures. We deduce from (6.3.13) that

$$|\nu_j|(\{x : \|x\| > u\}) = O(\exp(-Ku^2)) \qquad (u \to +\infty).$$

Evidently ν_0 is concentrated on Λ and ν_j on $\Lambda \cap e_j^\perp$ $(j = 1, \ldots, n)$. From (6.3.15) we deduce exactly as in Lemma 6.3.1 that $\Pi_3 = 0$ and we have proved the first part of the theorem with $\Pi_1(t) = -(\beta, t)$ and $\Pi_2 = Q$. If $P = 0$, we deduce as in Lemma 6.3.2 that $Q = \nu_j = 0$ $(j = 1, \ldots, n)$ and the theorem is proved.

Corollary 6.3.1 Let $\hat{p}$ be an infinitely divisible characteristic function admitting an $(\alpha, 0, \mu)$-Lévy–Hinčin representation with a finite measure μ satisfying the following conditions:

(a) μ is concentrated on a Linnik set $\Lambda = \Lambda_1 + \cdots + \Lambda_n$, Λ_j being a finite set $(j = 2, \ldots, n)$;
(b) for some $K > 0$

$$\mu(\{x : \|x\| > u\}) = O(\exp(-Ku^2)) \qquad (u \to +\infty).$$

If q divides p, then $\hat{q}$ admits a $(\beta, 0, \nu)$-Lévy–Hinčin representation where ν is a finite signed measure concentrated on Λ and satisfies

$$|\nu|(\{x : \|x\| > u\}) = O(\exp(-Ku^2)) \qquad (u \to +\infty).$$

Proof If Λ is constructed on the canonical basis, this follows directly from the preceding theorem, and if Λ is constructed on an arbitrary basis, this follows by isomorphism from this particular case.

6.4 SOME SUFFICIENT CONDITIONS FOR MEMBERSHIP TO I_0^n

We introduce now the following definition. Let μ be a signed measure concentrated on a Linnik set Λ constructed on $\{\theta_1, \ldots, \theta_n\}$ and let $\xi = \sum_{j=1}^n \xi_j \theta_j$ be a point of Λ $(\xi \neq 0)$. If $\xi_j \neq 0$ $(j = 1, \ldots, n)$, we say that ξ has an issue with respect to μ if there exist some k, $\omega_1, \ldots, \widehat{\omega_k}, \ldots, \omega_n$ $(1 \leq k \leq n,\ \omega_j = \pm 1)$ such that

$$|\mu|(\{x = \sum_{j=1}^n x_j \theta_j : x_k = \xi_k,\ \omega_l(x_l - \xi_l) > 0 \text{ for some } l \neq k\}) = 0. \tag{6.4.1}$$

If $\xi_{j_1} = \cdots = \xi_{j_\kappa} = 0$, we say that ξ has an *issue* with respect to μ if there exist some k, $\omega_1, \ldots, \widehat{\omega_k}, \ldots, \omega_n$ $(1 \leq k \leq n,\ k \neq j_1, \ldots, k \neq j_l,\ \omega_j = \pm 1)$ such that (6.4.1) is satisfied.

Theorem 6.4.1 Let $\hat{p}$ and $\hat{q}$ be some entire characteristic functions. We suppose that $\hat{p}$ admits for any $t \in C^n$ the representation

$$\log \hat{p}(t) = i(\alpha, t) - P(t) + \int (e^{i(t, u)} - 1)\mu(du) \tag{6.4.2}$$

where $\alpha = (\alpha_1, \ldots, \alpha_n) \in R^n$, P is a nonnegative quadratic form, and $\mu \in \mathscr{M}_n$ is concentrated on a Linnik set Λ constructed on the canonical basis of R^n and satisfies

$$|\mu|(\{x : \|x\| > u\}) = O(\exp(-Ku^2)) \qquad (u \to +\infty) \tag{6.4.3}$$

for some $K > 0$ and that $\hat{q}$ admits for any $t \in C^n$ the representation

$$\begin{aligned} \log \hat{q}(t) = {} & i(\beta, t) - Q(t) + \int (e^{i(t, u)} - 1)\nu_0(du) \\ & + i\sum_{j=1}^{n} t_j \int e^{i(t, u)}\nu_j(du) \end{aligned} \tag{6.4.4}$$

where $\beta = (\beta_1, \ldots, \beta_n) \in R^n$, Q is a nonnegative quadratic form, and $\nu_j \in \mathscr{M}_n$ are concentrated, respectively, on Λ $(j = 0)$ and on $\Lambda \cap e_j^{\perp}$ $(j = 1, \ldots, n)$. If q divides p and if a point $\xi \in \Lambda$ has an issue with respect to μ, then

$$\nu_j(\{\xi\}) = 0 \qquad (j = 1, \ldots, n), \qquad 0 \le \nu_0(\{\xi\}) \le \mu(\{\xi\}).$$

Proof From (6.4.2), we deduce that

$$\log \hat{p}(t) = a + i(\alpha, t) - P(t) + \sum_{\lambda_{k,l} \in \Lambda_k} L_l(t) e^{i\lambda_{k,l} t_k}$$

where $a \in R$ and

$$L_l(t) = \sum_{\substack{\{m = (m_1, \ldots, m_n): \\ m_k = \lambda_{k,l}\}}} \mu(\{m\}) \exp\left(i \sum_{j \neq k} m_j t_j\right).$$

and from (6.4.4), we obtain

$$\log \hat{q}(t) = b + i(\beta, t) + it_k \hat{\nu}_k(t) - Q(t) + \sum_{\lambda_{k,l} \in \Lambda_k} M_l(t) e^{i\lambda_{k,l} t_k}$$

where $b \in R$ and

$$M_l(t) = \sum_{\substack{\{m = (m_1, \ldots, m_n): \\ m_k = \lambda_{k,l}\}}} \left(\nu_0(\{m\}) + i \sum_{j \neq k} t_j \nu_j(\{m\})\right) \exp\left(i \sum_{j \neq k} m_j t_j\right).$$

From (6.4.3), we deduce easily the existence of some positive constant C such that

$$|\log \hat{p}(t)| = O(\|t\|^2 \exp(C\|\mathrm{Im}\, t\|^2)) \qquad (\|t\| \to +\infty)$$

and this relation and Theorem 5.3.3 implies

$$|\log \hat{q}(t)| = O(\|t\|^3 \exp(C\|\mathrm{Im}\, t\|^2)) \qquad (\|t\| \to +\infty).$$

From this and from

$$M_l(t) = \lim_{T\to\infty} (1/2T) \int_{-T}^{T} \log \hat{q}(t) e^{-i\lambda_{k,l} t_k}\, dt_k,$$

we deduce easily that

$$M_l(t) = O(\|t\|^3 \exp(C\|\mathrm{Im}\, t\|^2)) \qquad (\|t\| \to +\infty). \tag{6.4.5}$$

Now we define p_y and q_y by

$$\hat{p}_y(t) = \hat{p}(t + iy)/\hat{p}(iy), \qquad \hat{q}_y(t) = \hat{q}(t + iy)/\hat{q}(iy)$$

for any $y \in R^n$. If q divides p, from Corollaries 5.2.2 and 2.5.8, we deduce that $\mathrm{pr}_{e_k} q_y$ divides $\mathrm{pr}_{e_k} p_y$ for any $y \in R^n$. If $y \perp e_k$ and $t = (t_1, \ldots, t_n) \in R^n$

$$\log \mathrm{pr}_{e_k} \hat{p}_y(t) = i\alpha_k t_k - \gamma t_k^2 + \sum_{\lambda_{k,l} \in \Lambda_k} L_l(iy)(e^{i\lambda_{k,l} t_k} - 1),$$

$$\log \mathrm{pr}_{e_k} \hat{q}_y(t) = i(\beta_k + \hat{v}_k(iy))t_k - \delta(t_k^2) + \sum_{\lambda_{k,l} \in \Lambda_k} M_l(iy)(e^{i\lambda_{k,l} t_k} - 1).$$

Since $\mathrm{pr}_{e_k} p_y$ satisfies the assumptions of Corollary 6.1.1, we obtain from this result the inequalities

$$0 \le M_l(iy) \le L_l(iy) \tag{6.4.6}$$

for any $y \perp e_k$.

Let now l_0 be such that $\xi_k = \lambda_{k,l_0}$. We have

$$L_{k0}(iy) = O(e^{-(\xi, y)})$$

when

$$y \perp e_k, \qquad \omega_j y_j > 0 \quad (j \ne k), \qquad \|y\| \to +\infty. \tag{6.4.7}$$

From (6.4.6), we deduce that

$$M_{l_0}(iy) = O(e^{-(\xi, y)}) \tag{6.4.8}$$

holds under conditions (6.4.7). Letting $y_k = 0$, $\omega_j y_j > 0$ $(j \ne k, j \ne l)$ and

$$M_{l_0}(iy) = f(y_l), \tag{6.4.9}$$

we obtain from (6.4.5)

$$|f(y_l)| = O(|1 + y_l|^3 \exp(C(\mathrm{Re}\, y)^2))$$

when $|y_l| \to +\infty$ and from (6.4.8)

$$|f(y_l)| = O(e^{\xi_l y_l})$$

for $\operatorname{Im} y_l = 0$, $\omega_l \operatorname{Re} y_l \geq 0$. It is clear that

$$f(y_l) = \begin{cases} O(1) & \text{if} \quad \xi_l \neq 0, \\ O(|y_l|) & \text{if} \quad \xi_l = 0, \end{cases}$$

for $\operatorname{Re} y_l = 0$, $|\operatorname{Im} y_l| \to +\infty$.

Now we use the following lemma [cf. Lukacs (1970, Lemma 9.2.3)].

Lemma 6.4.1 If f is regular in the half-plane $\operatorname{Re} z \geq 0$ and satisfies

$$|f(z)| \leq M_1 |z+1|^a \quad \text{for} \quad \operatorname{Re} z = 0,$$

$$|f(z)| \leq M_2 e^{bz}(z+1)^c \quad \text{for} \quad \operatorname{Im} z = 0,$$

$$|f(z)| \leq M_3 \exp((d \operatorname{Re} z)^2 |z+1|^c) \quad \text{for} \quad \operatorname{Re} z \geq 0,$$

where M_j are nonnegative constants ($j = 1, 2, 3$) while a, b, c, and d are nonnegative constants with $c \geq a$, then

$$|f(z)| \leq M_1 |z+1|^a \exp(b \operatorname{Re} z)$$

in the half-plane $\operatorname{Re} z \geq 0$.

If $\xi_l \geq 0$, applying this lemma to the function defined by (6.4.9), we obtain

$$|f(y_l)| = O(|1 + y_l| e^{\xi_{l} y_l}) \tag{6.4.10}$$

when $|y_l| \to +\infty$. If $\xi_l < 0$, we apply the lemma to $f(y_l)e^{-\xi_{l} y_l}$ and obtain

$$|f(y_l)| = O(e^{\xi_{l} y_l}) \tag{6.4.11}$$

when $|y_l| \to +\infty$. Noting that

$$f(y_l) = \sum_{\lambda_{l,\alpha} \in \Lambda_l} M_{l,\alpha}(y) e^{-\lambda_{l,\alpha} y_l}$$

where

$$M_{l,\alpha}(y) = \sum_{\substack{\{m=(m_1, \ldots, m_n):\\ m_k = \xi_k,\, m_l = \lambda_{l,\alpha}\}}} \Big(v_0(\{m\}) - \sum_{\substack{j \neq k \\ j \neq l}} y_j v_j(\{m\})\Big) \exp\Big(-\sum_{\substack{j \neq k \\ j \neq l}} m_j y_j\Big),$$

we deduce from (6.4.10), (6.4.11), and Theorem 3.6.1 that

$$M_{l,\alpha}(y) = 0$$

for any (l, α) satisfying $\omega_l(\lambda_{l,\alpha} - \xi_l) > 0$ and $\omega_j y_j > 0$ ($j \neq k$, $j \neq l$). Since $M_{l,\alpha}(y)$ is an entire function of $y_1, \ldots, y_n$ independent of y_k and y_l, we obtain

$$M_{l,\alpha} = 0$$

for any α and j_l satisfying $\omega_l(\lambda_{l,\alpha} - \xi_l) > 0$. From this, we deduce easily that

$$v_j(\{m\}) = 0$$

$(j = 0, 1, \ldots, n)$ for any $m = (m_1, \ldots, m_n)$ satisfying $m_k = \xi_k$, $\omega_l(m_l - \xi_l) > 0$. Since this is true for any $l \neq k$, we obtain

$$\nu_j\left(\left\{x = \sum_{j=1}^{n} x_j e_j : x_k = \xi_k, \omega_l(x_l - \xi_l) > 0 \text{ for some } l \neq k\right\}\right) = 0$$

$(j = 0, 1, \ldots, n)$.

Now we obtain easily that

$$L_{k0}(y) = (\mu(\{\xi\}) + o(1))e^{-(\xi, y)}$$

and

$$M_{k0}(y) = \left(\nu_0(\{\xi\}) - \sum_{j \neq k} y_j \nu_j(\{\xi\}) + o(1)\right)e^{-(\xi, y)}$$

when $\omega_j y_j > 0$, $\|y\| \to +\infty$ $(j = 1, \ldots, n; j \neq k)$. From (6.4.6), we deduce that

$$\nu_j(\{\xi\}) = 0$$

for any $j \neq k$ and

$$0 \leq \nu_0(\{\xi\}) \leq \mu(\{\xi\}).$$

Finally, since $\xi_k \neq 0$, we have

$$\nu_k(\{\xi\}) = 0$$

and the theorem is proved.

With the same method and some supplementary technical difficulties, we have the following result (we omit the proof).

Theorem 6.4.2 The conclusions of Theorem 6.4.1 are true again if we replace (6.4.2) by

$$\log \hat{p}(t) = i(\alpha, t) - P(t) + \int K(t, u)\mu(du)$$

and (6.4.4) by

$$\log \hat{q}(t) = i(\beta, t) - Q(t) + \int K(t, u)\nu_0(du) + i\sum_{j=1}^{n} t_j \int K(t, u)\nu_j(du),$$

the α, P, μ, β, Q, and ν_j $(0 \leq j \leq n)$ satisfying the same conditions.

We have then

Theorem 6.4.3 Let $\hat{p}$ be a characteristic function admitting an (α, P, μ)-Lévy–Hinčin representation. If $\mu \in \mathscr{M}_n$ satisfies the following conditions:

(a) μ is concentrated on a Linnik set $\Lambda = \Lambda_1 + \cdots + \Lambda_n$, Λ_j having no finite accumulation point ($j = 1, \ldots, n$);
(b) each point of Λ has an issue with respect to μ;
(c) for some $K > 0$,

$$|\mu|(\{x : \|x\| > u\}) = O(\exp(-Ku^2))$$

when $u \to +\infty$,

then p belongs to I_0^n.

Proof We can suppose without loss of generality that $\{\theta_1, \ldots, \theta_n\}$ is the canonical basis of R^n, the general case following from this one by isomorphism. In this case, if q divides p, it follows from assumptions (a) and (c) and Theorem 6.3.2 that $\hat{q}$ admits for any $t \in C^n$ a representation (6.4.4) with $\beta \in R^n$, Q a nonnegative quadratic form and $\nu_j \in \mathscr{M}_n$ concentrated, respectively, on Λ ($j = 0$) and $\Lambda \cap e_j$ ($j = 1, \ldots, n$). It follows then from assumption (b) and Theorem 6.4.1 that

$$\nu_j(\{\xi\}) = 0$$

($j = 1, \ldots, n$) and

$$\nu_0(\{\xi\}) \geq 0$$

for any $\xi \in \Lambda$, so that q is infinitely divisible and p belongs to I_0^n.

With the same proof, using Theorems 6.3.3 and 6.4.2 instead of Theorems 6.3.2 and 6.4.1, we have also

Theorem 6.4.4 Let $\hat{p}$ be a characteristic function admitting an (α, P, μ)-Lévy–Hinčin representation. If $\mu \in \mathscr{M}_n$ satisfies the following conditions:

(a) μ is concentrated on a Linnik set $\Lambda = \Lambda_1 + \cdots + \Lambda_n$, Λ_j being a finite set ($j = 2, \ldots, n$);
(b) each point of Λ has an issue with respect to μ;
(c) for some $K > 0$,

$$|\mu|(\{x : \|x\| > u\}) = O(\exp(-Ku^2))$$

when $u \to +\infty$,

then p belongs to I_0^n.

Corollary 6.4.1 Let $\lambda_k = (\lambda_{k,1}, \ldots, \lambda_{k,n}) \in R^n$ ($k = 1, 2, \ldots$) be such that $\lambda_{k',j}/\lambda_{k,j}$ is either negative or an integer greater than one for $k' > k$ and

$j = 1, \ldots, n$ and $l_k = (l^{(k)}_{1, 0, \ldots, 0}, l^{(k)}_{0, 1, 0, \ldots, 0}, \ldots, l^{(k)}_{1, 1, \ldots, 1}) \in (R^+)^{2^n - 1}$ be such that

$$\sum_{k=1}^{\infty} \left(\sum_{\varepsilon} l^{(k)}_{\varepsilon_1, \ldots, \varepsilon_n} \right) < +\infty$$

and

$$\sum_{\{|\lambda_k| > u\}} \left(\sum_{\varepsilon} l^{(k)}_{\varepsilon_1, \ldots, \varepsilon_n} \right) = O(\exp(-Ku^2))$$

for some $K > 0$. Then the sequence of probabilities $\{q_k\}$ defined by

$$q_k = \mathfrak{n}(\alpha, P) * \mathfrak{p}(\lambda_1, l_1) * \cdots * \mathfrak{p}(\lambda_k, l_k)$$

converges completely to a probability p belonging to I_0^n.

Proof From the assumptions, we deduce easily that $\hat{q}_k$ converges to a function f uniformly on any compact set of R^n. From Lévy's theorem, we deduce the existence of a probability p such that $\hat{p} = f$. From Theorem 4.3.5 we deduce that p satisfies the conditions of the preceding theorem and therefore belongs to I_0^n.

We prove now the existence of convolutions of two Poisson-type probabilities which do not belong to I_0^n. From a result due to Lévy [cf. Ramachandran (1967, p. 122)], this is impossible in the case $n = 1$. We give an example in the case $n \geq 2$. Let p be the probability defined by

$$\log \hat{p}(t) = \sum_{j=1}^{6} l_j e^{i(a_j, t)}$$

where $l_j > 0$ $(j = 1, \ldots, 6)$, $a_1 = e_1$, $a_2 = 2e_2$, $a_3 = e_1 + 2e_2$, $a_4 = 2e_1$, $a_5 = e_2$, $a_6 = 2e_1 + e_2$. Then p is the convolution of two Poisson-type probabilities with respective Poisson spectra $\{e_1, 2e_2, e_1 + 2e_2\}$ and $\{2e_1, e_2, 2e_1 + e_2\}$. Now $p = \exp(-\mu(R^n)) \exp \mu$ with

$$\mu(B) = \sum_{a_j \in B} l_j$$

for any $B \in \mathscr{B}_n$. If we consider the signed measure $\nu = \mu - \varepsilon\delta_{e_1 + e_2}$, then we prove easily the existence of some $\varepsilon_0 > 0$ such that ν^{*2}, ν^{*3}, and $\nu + \frac{1}{2}\nu^{*2}$ are measures if $0 < \varepsilon < \varepsilon_0$. Since for any integer $k \geq 2$, there exist nonnegative integers r and s such that $k = 2r + 3s$, ν^{*k} is a measure if $0 < \varepsilon < \varepsilon_0$ and $k \geq 2$. This implies that $\exp \nu$ is a measure if $0 < \varepsilon < \varepsilon_0$. Then the probability $q = \exp(-\nu(R^n)) \exp \nu$ divides p and is not infinitely divisible, so that p does not belong to I_0^n. It must be noted that p does not belong to I_0^n since $e_1 + e_2$ has no issue with respect to μ, while $e_1 + e_2$ does not belong to $S(\mu)$. Therefore, condition (b) in Theorems 6.4.3 and 6.4.4 must be satisfied for any point of Λ (and not only for the points of the Poisson spectrum of p).

NOTES

For the case $n = 1$, the study of infinitely divisible probabilities with normal factor was begun by Linnik (1958, 1959a,b, 1960). He proved the necessary condition (Theorem 6.1.1) and also some sufficient conditions. These results were extended by Ostrovskiĭ (1965c) who proved Theorem 6.1.2.

For the case $n > 1$, the first study was due to Cuppens (1967, 1969a) who proved essentially Corollary 6.4.1. Later, Livsič (1970) introduced for the case $n = 2$ the notion of issue and proved in this case a result which is stronger than Theorem 6.4.4. The results given in Sections 6.2–6.4 are new while the ideas are those of Cuppens (1967). The example of a convolution of two Poisson probabilities having indecomposable factor was due to Cuppens (1969c).

Chapter 7

Infinitely Divisible Probabilities without Normal Factor

In this chapter, we study the infinitely divisible probabilities without normal factor. In the first section, we give some sufficient conditions for such a probability to belong to I_0^n. In the second section, we prove that one of this condition is also necessary for probabilities having absolutely continuous Poisson measures. Finally we study the probabilities having Poisson measures concentrated on $(N)A$, A being an independent set.

7.1 PROBABILITIES WITH A POISSON MEASURE CONCENTRATED ON A STRIP

Theorem 7.1.1 Let $\hat{p}$ be an n-variate characteristic function admitting an $(\alpha, 0, \mu)$-De Finetti representation where $\mu \in \mathcal{M}_n$ satisfies

$$\|\mu\| < \log 2. \tag{7.1.1}$$

If q divides p, then $\hat{q}$ admits a $(\beta, 0, \nu)$-De Finetti representation with a finite signed measure ν satisfying

$$\nu(R^n) < \log 2. \tag{7.1.2}$$

Moreover, if μ is concentrated on $A \in F_\sigma^n$, then ν is concentrated on $C(\mu) \cap (N)A$ ($C(\mu)$ is the convex support of μ) and $\nu(B) \geq 0$ for any Borel set B satisfying

$$B \cap \bigcup_{k=2}^{\infty} (k)A = \varnothing. \tag{7.1.3}$$

Proof We can suppose without any loss of generality that p is infinitely divisible. Indeed, if p is not infinitely divisible, we consider the characteristic function $\hat{r}$ admitting the $(\alpha, 0, |\mu|)$-De Finetti representation and it is clear that q divides the infinitely divisible probability r.

If p is infinitely divisible, μ is a measure. From

$$p = \frac{\delta_\alpha * \exp \mu}{\exp(\mu(R^n))}$$

and from (7.1.1) we deduce that

$$p(\{\alpha\}) \geq \exp(-\|\mu\|) > \tfrac{1}{2}.$$

If q divides p, there exists some $\beta \in R^n$ such that

$$q(\{\beta\}) > \tfrac{1}{2}$$

and, from Theorem 4.3.7, $\hat{q}$ admits a $(\beta, 0, \nu)$-De Finetti representation with a finite signed measure ν satisfying (7.1.2).

If μ is concentrated on A, then p is concentrated on $(N)A + \alpha$ and q is concentrated on $(N)A + \beta$. From Theorem 4.3.7, ν is concentrated on $(N)A$ and if B is a Borel set satisfying (7.1.3), then $\nu(B) \geq 0$.

Now, let h be the support function of $C(\mu)$. If for some θ, $h(\theta) \geq 0$, it follows from Corollary 5.2.5 that

$$\operatorname{ext}_\theta(\nu) \leq h(\theta). \tag{7.1.4}$$

If $h(\theta) < 0$, then

$$A \subset S = \{x \in R^n : (x, -\theta) \geq -h(-\theta)\}$$

and

$$\bigcup_{k=1}^{\infty} (k)A \subset S,$$

so that (7.1.4) holds also in this case. Since

$$C(\mu) = \bigcap_{\theta \in R^n \setminus \{0\}} \Pi_\theta$$

where Π_θ is the half-space $\{x \in R^n : (x, \theta) \leq h(\theta)\}$ and since $C(\nu)$ is the smallest closed convex set on which ν is concentrated, we find that

$$C(\nu) \subset C(\mu)$$

so that ν is concentrated on $C(\mu)$ and the theorem is proved.

Theorem 7.1.2 Let $\hat{p}$ be a characteristic function admitting an $(\alpha, 0, \mu)$-De Finetti representation with a finite signed measure μ concentrated on a subset $A \in F_\sigma^n$ of the strip $S = \{x \in R^n : 0 \leq a \leq \mathrm{pr}_\theta\, x \leq b < +\infty\}$ for some $\theta \in R^n$. If q divides p, then $\hat{q}$ admits a $(\beta, 0, v)$-De Finetti representation with a finite signed measure v concentrated on $(N)A \cap C(\mu)$. Moreover $v(B) \geq 0$ for any Borel set B satisfying (7.1.3).

Proof We can suppose again without any loss of generality that p is infinitely divisible. If Γ is the one-dimensional subspace generated by θ, p belongs to $\mathscr{A}_\Gamma$ and, from Corollary 5.2.1, $q \in \mathscr{A}_\Gamma$. From Corollary 5.2.2, we deduce that the functions $\hat{p}_l$ and $\hat{q}_l$ defined by

$$\hat{p}_l(t) = \hat{p}(t + il\theta)/\hat{p}(il\theta) \qquad \text{and} \qquad \hat{q}_l(t) = \hat{q}(t + il\theta)/\hat{q}(il\theta)$$

are characteristic functions and that q_l divides p_l for any $l \in R$.

Since $\mathrm{pr}_\theta\, \hat{\mu}$ is an entire function, we obtain easily that $\hat{p}_l$ admits an $(\alpha, 0, \mu_l)$-De Finetti representation where

$$\mu_l(B) = \int_B e^{-(l\theta, u)} \mu(du)$$

for any $B \in \mathscr{B}_n$. Since

$$\lim_{l \to \infty} \mu_l(R^n) = 0,$$

we can choose $m \in R^n$ such that

$$\mu_m(R^n) < \log 2.$$

Then p_m satisfies all the assumptions of Theorem 7.1.1 and from this theorem we deduce that $\hat{q}_m$ admits a $(\beta, 0, v_m)$-De Finetti representation with a finite signed measure v_m concentrated on $(N)A \cap C(\mu)$ and such that $v_m(B) \geq 0$ for any Borel set B satisfying (7.1.3). If we define v by

$$v(B) = \int_B e^{(m, u)} v_m(du),$$

since $C(\mu) \subset S$, v is a finite signed measure concentrated on $(N)A \cap C(\mu)$ such that $v(B) \geq 0$ for any Borel set B satisfying (7.1.3). It is easily seen that $\hat{q}$ admits a $(\beta, 0, v)$-De Finetti representation.

Corollary 7.1.1 If $\hat{p}$ admits an $(\alpha, 0, \mu)$-Lévy–Hinčin representation with a measure μ concentrated on $A = \{x \in R^n : a < \mathrm{pr}_\theta\, x < 2a\}$ for some $\theta \in R^n$ and some $a > 0$, then p belongs to I_0^n.

Proof It is clear that $0 \notin S(\mu)$ and, from Corollary 4.3.1, we deduce that $\hat{p}$ admits an $(\alpha', 0, \mu')$-De Finetti representation with a measure μ' concentrated on A. If q divides p, from the preceding theorem, we deduce that $\hat{q}$ admits a $(\beta, 0, v)$-De Finetti representation with $v \in \mathscr{M}_n$ concentrated on

$C(\mu) \cap (N)A$. Since $C(\mu) \subset \{x \in R^n : a \le \mathrm{pr}_\theta x \le 2a\}$ and $\bigcup_{k=2}^{\infty} (k)A \subset \{x \in R^n : 2a < \mathrm{pr}_\theta x\}$, ν is concentrated on A. From Theorem 7.1.2, we deduce then that $\nu(B) \ge 0$ for any Borel set B and q is infinitely divisible, so that $p \in I_0^n$.

Corollary 7.1.2 If $A \subset R^n$ is an open convex set satisfying $A \cap (2)A = \varnothing$ and if $\hat{p}$ admits an $(\alpha, 0, \mu)$-Lévy–Hinčin representation with a measure μ concentrated on A, then p belongs to I_0^n.

Proof If A is an open convex set satisfying $A \cap (2)A = \varnothing$, there exist some $\theta \in R^n$ and some $a > 0$ such that A is contained in $\{x \in R^n : a < \mathrm{pr}_\theta x < 2a\}$ and the conclusion follows from the preceding corollary.

From Corollary 7.1.1, we obtain

Theorem 7.1.3 If p is an infinitely divisible probability, there exists a sequence $\{q_k\}$ of probabilities without indecomposable factor such that

$$\lim_{k \to \infty} (q_0 * \cdots * q_k) \stackrel{c}{=} p.$$

Proof If p is infinitely divisible, $\hat{p}$ admits an (α, Q, μ)-Lévy–Hinčin representation. For some $\theta \in R^n$, we factor p as

$$p = p_0 * p_1 * p_2$$

where p_0 is the normal probability with $(\alpha, Q, 0)$-Lévy–Hinčin representation and $\hat{p}_j$ $(j = 1, 2)$ admits a $(0, 0, \mu_j)$-Lévy–Hinčin representation where

$$\mu_1(B) = \mu(B \cap \{x \in R^n : \mathrm{pr}_\theta(x) \ne 0\}),$$
$$\mu_2(B) = \mu(B \cap \{x \in R^n : \mathrm{pr}_\theta(x) = 0\})$$

for any $B \in \mathscr{B}_n$. If we consider the sequence $\{\hat{q}_{1,k}\}$ of characteristic functions admitting a $(0, 0, \nu_{1,k})$-Lévy–Hinčin representation with

$$\nu_{1,k}(B) = \mu_1(B \cap A_k)$$

for any $B \in \mathscr{B}_n$ where A_k is defined by

$$A_{4m+1} = \{x \in R^n : \alpha^m \le \mathrm{pr}_\theta\, x < \alpha^{m+1}\},$$
$$A_{4m+2} = \{x \in R^n : \alpha^{-m-1} \le \mathrm{pr}_\theta\, x < \alpha^{-m}\},$$
$$A_{4m+3} = \{x \in R^n : -\alpha^{m+1} \le \mathrm{pr}_\theta\, x < -\alpha^{m}\},$$
$$A_{4m+4} = \{x \in R^n : -\alpha^{-m} \le \mathrm{pr}_\theta\, x < -\alpha^{-m-1}\}$$

for some $\alpha \in R$, $1 < \alpha < 2$. It follows then from Corollary 7.1.1 that $q_{1,k} \in I_0^n$ $(k \ge 1)$ and from Theorem 4.3.5 that

$$\lim_{k \to \infty} (q_{1,1} * \cdots * q_{1,k}) \stackrel{c}{=} p_1.$$

In the same way, we can prove the existence of a sequence $\{q_{2,k}\}$ of probabilities without indecomposable factor such that

$$\lim_{k\to\infty} (q_{2,1} * \cdots * q_{2,k}) \stackrel{c}{=} p_2 .$$

Since p_0 has no indecomposable factor, the theorem is proved.

We give now an extension of Corollary 7.1.2.

Theorem 7.1.4 Let A be an open set satisfying

$$A^* \cap \bigcup_{k=2}^{\infty} (k)A = \varnothing \tag{7.1.5}$$

where A^* is the convex hull of A. If $\hat{p}$ is a characteristic function admitting an $(\alpha, 0, \mu)$-Lévy–Hinčin representation with a finite measure μ concentrated on A, then p belongs to I_0^n.

Proof If A satisfies (7.1.5), then $0 \notin \overline{A}$. Indeed, if $0 \in \overline{A}$, there exists a sequence $\{a_k\}$ of elements of A such that

$$\lim_{k\to\infty} a_k = 0.$$

If $b \in A$, then $a_k + b \in (2)A$ and

$$\lim_{k\to\infty} (a_k + b) = b$$

and, since A is open, $a_k + b$ belongs to A if k is great enough, so that $0 \in A \cap (2)A$. From Corollary 4.3.1, we deduce then that $\hat{p}$ admits an $(\alpha', 0, \mu')$-De Finetti representation with a measure μ' concentrated on A.

If $0 \in A^*$, then there exists some $\alpha_k > 0$ and $a_k \in A$ $(k = 1, \ldots, p)$ such that

$$\sum_{k=1}^{p} \alpha_k a_k = 0.$$

When A is open, we can suppose that the α_k are positive rationals $(k = 1, \ldots, p)$ and this implies easily that

$$\sum_{k=1}^{p} m_k a_k = 0$$

for $m_k \in N$, so that $0 \in \bigcup_{k=2}^{\infty} (k)A$. Therefore if A is an open set satisfying (7.1.5), $0 \notin A^*$. Since $0 \notin A^*$, from the properties of convex sets, we deduce the existence of some $\theta \in R^n$ such that $A^* \subset \{x \in R^n : \mathrm{pr}_\theta\, x \geq 0\}$. If $\Gamma = \{x \in R^n : x = l\theta,\ l \geq 0\}$, then p belongs to $\mathscr{A}_\Gamma$ and, if q divides p, from Corollary 5.2.2, we deduce that the functions $\hat{p}_l$ and $\hat{q}_l$ defined by

$$\hat{p}_l(t) = \hat{p}(t + il\theta)/\hat{p}(il\theta) \qquad \text{and} \qquad \hat{q}_l(t) = \hat{q}(t + il\theta)/\hat{q}(il\theta)$$

are characteristic functions and that q_l divides p_l for any $l \geq 0$.

Since p is infinitely divisible, we obtain easily that $\hat{p}_l$ admits an $(\alpha', 0, \mu_l)$-De Finetti representation with

$$\mu_l(B) = \int_B e^{-(l\vartheta, u)} \mu'(du)$$

for any $B \in \mathscr{B}_n$. Since

$$\lim_{l \to \infty} \mu_l(R^n) = 0,$$

we can choose some $m \in R^n$ such that

$$\mu_m(R^n) < \log 2.$$

Then p_m satisfies all the assumptions of Theorem 7.1.1 and from this theorem we deduce that $\hat{q}_m$ admits a $(\beta, 0, \nu_m)$-De Finetti representation with a finite signed measure ν_m concentrated on $(N)A \cap \overline{A^*}$ and such that $\nu_m(B) \geq 0$ for any Borel set B satisfying (7.1.3). Since A^* and $\bigcup_{k=2}^{\infty} (k)A$ are open sets, we deduce from (7.1.5) that ν_m is nonnegative and q_m is infinitely divisible. This implies that $p_m \in I_0^n$, so that p belongs to I_0^n.

If A is an open convex set, then

$$A \cap (2)A = \varnothing$$

is equivalent to (7.1.5), so that Theorem 7.1.4 implies Corollary 7.1.2. But there exist open sets A satisfying (7.1.5) and

$$A^* \cap (2)A^* \neq \varnothing. \tag{7.1.6}$$

We give now such an example for the case $n = 2$.

Let $C \subset R^2$ be the set composed of the two segments joining the points $2e_2$ and $2e_1 + 5e_2$ and the points $2e_1 + 5e_2$ and $4e_1 + 2e_2$. Then $(2)C$ is composed of the parallelogram with vertices $2e_1 + 7e_2$, $4e_1 + 10e_2$, $6e_1 + 7e_2$, and $4e_1 + 4e_2$ and the two segments joining the points $4e_2$ and $2e_1 + 7e_2$ and the points $6e_1 + 7e_2$ and $8e_1 + 4e_2$. If C^* is the convex hull of C, then $C^* \cap (2)C = \varnothing$ and since $\bigcup_{k=3}^{\infty} (k)C \subset \{x \in R^2 : \mathrm{pr}_{e_2} x \geq 6\}$, the distance between C^* and $\bigcup_{k=2}^{\infty} (k)C$ is positive. If $A = \{x \in R^2 : \|x - y\| < \varepsilon$ for some $y \in C\}$, then A is open and if ε is small enough, A satisfies (7.1.5) and (7.1.6).

We give now some extensions of the preceding results, combining them and the notion of Linnik sets introduced in Chapter 6. In the case $n = 1$, we have

Theorem 7.1.5 Let $\Lambda = \{\lambda_k\}$ be a Linnik set of real numbers and a and b be some real numbers satisfying

$$\cdots < \lambda_{-2} < \lambda_{-1} < a \leq 0 = \lambda_0 \leq b < \lambda_1 < \lambda_2 < \cdots.$$

We suppose that the finite signed measure μ is concentrated on $\Lambda \cup [a, b]$ and satisfies

$$\|\mu\| < \log 2$$

and for some $K > 0$

$$|\mu|(\{x : \|x\| > u\}) = O(\exp(-Ku^2))$$

when $u \to +\infty$. If the univariate characteristic function $\hat{p}$ admits an $(\alpha, 0, \mu)$-De Finetti representation and if q divides p, then $\hat{q}$ admits a $(\beta, 0, \nu)$-De Finetti representation with a finite signed measure ν concentrated on $\Lambda \cup [a, b]$ and satisfying

$$0 \leq \nu(\{\lambda\}) \leq \mu(\{\lambda\})$$

for any $\lambda \in \Lambda$.

Proof If we define ϕ by

$$\phi(z) = \alpha z + \int_a^b (e^{zu} - 1)\mu(du) - \mu(\Lambda),$$

ϕ is an entire function which is real for real z and satisfies

$$\phi(z) = \begin{cases} O(|z| \exp(b \operatorname{Re} z)) & \text{for} \quad \operatorname{Re} z \geq 0, \\ O(|z| \exp(a \operatorname{Re} z)) & \text{for} \quad \operatorname{Re} z \leq 0 \end{cases}$$

when $|z| \to +\infty$. Moreover, we have the representation

$$\log \hat{p}(t) = \phi(it) + \sum_{k=-\infty}^{+\infty} l_k \exp(i\lambda_k t)$$

for any $t \in C$ where $l_k = \mu(\{\lambda_k\})$. By assumption

$$\sum_{\{\lambda_k > u\}} l_k = O(\exp(-Ku^2))$$

when $u \to +\infty$ and from the ridge property $l_k \geq 0$ $(k = 1, 2, \ldots)$.

If q divides p, we deduce then from Theorem 9.5.1 of Lukacs (1970) that

$$\log \hat{q}(t) = \psi(it) + \sum_{k=-\infty}^{+\infty} m_k \exp(i\lambda_k t) \tag{7.1.7}$$

for any $t \in C$ where $0 \leq m_k \leq l_k$ $(k = 1, 2, \ldots)$ and ψ is an entire function which is real for real z and satisfies the estimates

$$\psi(z) = \begin{cases} O(|z|^3 \exp(b \operatorname{Re} z)) & \text{for} \quad \operatorname{Re} z \geq 0, \\ O(|z|^3 \exp(a \operatorname{Re} z)) & \text{for} \quad \operatorname{Re} z \leq 0 \end{cases} \tag{7.1.8}$$

when $|z| \to +\infty$. Now, from Theorem 7.1.1, we have

$$\log \hat{q}(t) = i\beta t + \int (e^{itu} - 1)\nu(du) \tag{7.1.9}$$

for any $t \in R$. Comparing (7.1.7) and (7.1.9), we have

$$g(it) = \psi(it) - \int_a^b e^{itu}\nu(du) + \nu(R)$$
$$= \int_{\{u \notin [a, b]\}} e^{itu}\nu(du) - \sum_{k=-\infty}^{+\infty} m_k \exp(i\lambda_k t) = \hat{\nu}^*(t)$$

where ν^* is a finite signed measure concentrated on $[a, b]^c$. From (7.1.8), we deduce that

$$g(z) = \begin{cases} O(|z|^3 \exp(b \operatorname{Re} z)) & \text{for } \operatorname{Re} z \geq 0, \\ O(|z|^3 \exp(a \operatorname{Re} z)) & \text{for } \operatorname{Re} z \leq 0 \end{cases}$$

when $|z| \to +\infty$. From Theorem 3.6.1, we obtain that ν^* is concentrated on $[a, b]$, so that $\nu^* = g = 0$. From this, we deduce easily the theorem.

We do not know if such a result holds for multivariate probabilities. Nevertheless, we have

Theorem 7.1.6 Let $\Lambda = \Lambda_1 + \cdots + \Lambda_n$ where $\Lambda_j = \{\lambda_{j,k} e_j\}$, $\{\lambda_{j,k}\}_{k=0}^{+\infty}$ being a Linnik set of positive numbers ($j = 1, \ldots, n$), and let a_j be a number such that

$$\lambda_{j,0} = 0 \leq a_j < \lambda_{j,1} < \lambda_{j,2} < \cdots \qquad (j = 1, \ldots, n).$$

We suppose that the finite signed measure μ is concentrated on $\Lambda \cup A$ ($A = \prod_{j=1}^n [0, a_j]$) and satisfies

$$\|\mu\| < \log 2$$

and for some $K > 0$

$$|\mu|(\{x : \|x\| > u\}) = O(\exp(-Ku^2))$$

when $u \to +\infty$. If the n-variate characteristic function $\hat{p}$ admits an $(\alpha, 0, \mu)$-De Finetti representation and if q divides p, then $\hat{q}$ admits a $(\beta, 0, \nu)$-De Finetti representation with ν concentrated on $\Lambda \cup A$. Moreover, if $\xi \in \Lambda$ has an issue with respect to μ, then

$$0 \leq \nu(\{\xi\}) \leq \mu(\{\xi\}). \tag{7.1.10}$$

Proof From Theorem 7.1.1, we have

$$\log \hat{q}(t) = i(\beta, t) + \int (e^{i(t, u)} - 1)\nu(du) \tag{7.1.11}$$

for any $t \in R^n$. Since ν is concentrated on $(R^+)^n$, (7.1.11) holds for any t such that $\operatorname{Im} t \geq 0$. Defining p_y, q_y, μ_y, and ν_y by

$$\begin{aligned} \hat{p}_y(t) &= \hat{p}(t + iy)/\hat{p}(iy), & \hat{q}_y(t) &= \hat{q}(t + iy)/\hat{q}(iy), \\ \hat{\mu}_y(t) &= \hat{\mu}(t + iy), & \hat{\nu}_y(t) &= \hat{\nu}(t + iy), \end{aligned} \tag{7.1.12}$$

we obtain from Corollaries 5.2.2 and 2.5.8 that $\operatorname{pr}_{e_j} q_y$ divides $\operatorname{pr}_{e_j} p_y$ for any $j = 1, \ldots, n$ and any $y \in R^n$. We have

$$\begin{aligned} \log \operatorname{pr}_{e_j} p_y(t) = i\alpha_j t_j &+ \int_{\{0 \leq (u, e_j) \leq a_j\}} (e^{it_j u_j} - 1) \operatorname{pr}_{e_j} \mu_y(du) \\ &+ \sum_{k=1}^{\infty} l_{j,k}(y)(e^{i\lambda_{j,k} t_j} - 1) \end{aligned}$$

with

$$l_{j,k}(y) = \sum_{\substack{\{x = (x_1, \ldots, x_n): \\ x_j = \lambda_{j,k}\}}} \mu_y(\{x\}),$$

so that $\operatorname{pr}_{e_j} p_y$ satisfies the assumptions of the preceding theorem. From (7.1.11), we obtain for $y \in (R^+)^n$

$$\log \operatorname{pr}_{e_j} q_y(t) = i\beta_j t_j + \int (e^{it_j u_j} - 1) \operatorname{pr}_{e_j} \nu_y(dy)$$

and, from Theorem 7.1.5, we deduce that $\operatorname{pr}_{e_j} \nu_y$ is concentrated on $[0, a_j] \cup \Lambda_j$ for any $y \in (R^+)^n$. Therefore

$$\int_B e^{-(y, u)} \nu(du) = 0$$

for any $y \in (R^+)^n$ and any B satisfying

$$B \subset L_{j,k} = \{\lambda_{j,k} < \operatorname{pr}_{e_j} u < \lambda_{j,k+1}\}$$
$$(j = 1, \ldots, n, \quad k = 1, 2, \ldots). \tag{7.1.13}$$

From this we deduce that $\nu(B) = 0$ for any B satisfying (7.1.13). Since this is true for any $j = 1, 2, \ldots, n$ and any $k = 1, 2, \ldots$, ν is concentrated on

$$(R^+)^n \cap \left(\bigcup_{j=1}^{n} \bigcup_{k=1}^{\infty} L_{j,k} \right)^c = \Lambda \cup A.$$

We have now

$$\log \hat{q}(t) = i(\beta, t) + \int_A (e^{i(t, u)} - 1)\nu(du) + \sum_{\lambda \in \Lambda} m_\lambda(e^{i(\lambda, t)} - 1) \tag{7.1.14}$$

for any $t \in R^n$ and since

$$t \to \log \hat{q}(t) - i(\beta, t) + \int_A (e^{i(t, u)} - 1)v(du)$$

is an entire function, we obtain that (7.1.14) holds for any $t \in C^n$. If $\xi \in \Lambda$ has an issue with respect to μ, with the method used in Theorem 6.4.1, we deduce from this inequalities (7.1.10).

Theorem 7.1.7 Let $\Lambda = \Lambda_1 + \cdots + \Lambda_n$ where $\Lambda_j = \{\lambda_{j,k} e_j\}$, $\{\lambda_{j,k}\}_{k=0}^{+\infty}$ being a Linnik set of positive numbers ($j = 1, \ldots, n$), and let a_j be a real number satisfying

$$\lambda_{j,0} = 0 \leq a_j < \lambda_{j,1} < \lambda_{j,2} < \cdots$$

($j = 1, \ldots, n$) and $A = \prod_{j=1}^n [0, a_j] \cap \{x : (x, \sum_{j=1}^n a_j e_j) > \frac{1}{2}\}$. We suppose that a finite signed measure μ satisfies the following conditions:

(a) μ is concentrated on $\Lambda \cup A$;
(b) each point of Λ has an issue with respect to μ;
(c) for some $K > 0$,

$$|\mu|(\{x : \|x\| > u\}) = O(\exp(-Ku^2))$$

when $u \to +\infty$.

If the characteristic function $\hat{p}$ admits an $(\alpha, 0, \mu)$-De Finetti representation, then p belongs to I_0^n.

Proof Let p_y, q_y, and μ_y be defined by (7.1.12). Then $\hat{p}_y$ admits an $(\alpha, 0, \mu_y)$-De Finetti representation and, if q divides p, q_y divides p_y for any $y \in R^n$. Since μ is concentrated on $(R^+)^n$, we can choose y so that

$$\|\mu_y\| < \log 2.$$

We deduce then from Theorem 7.1.6 that $\hat{q}_y$ admits a $(\beta, 0, v)$-De Finetti representation with v concentrated on $\Lambda \cup \prod_{j=1}^n [0, a_j]$ and satisfies $v(\{\lambda\}) \geq 0$ for any $\lambda \in \Lambda$. From Theorem 7.1.1, we deduce that v is also concentrated on $\{x : (x, \sum_{j=1}^n a_j e_j) > \frac{1}{2}\}$, so that v is concentrated on $\Lambda \cup A$.

Now, for any Borel set B, we have

$$(q_y * \delta_{-\beta})(B) = \sum_{k=0}^{\infty} v^{*k}(B).$$

If $B \subset A$, then $0 \notin B$ and $B \cap \bigcup_{k=2}^{\infty} (k)A = \varnothing$, so that

$$v(B) = (q_y * \delta_{-\beta})(B) \geq 0$$

and v is nonnegative. Therefore q_y is infinitely divisible and q is also infinitely divisible, so that $p \in I_0^n$.

7.2 PROBABILITIES HAVING AN ABSOLUTELY CONTINUOUS POISSON MEASURE

We want to prove

Theorem 7.2.1 Let $\hat{p}$ be a characteristic function admitting an $(\alpha, 0, \mu)$-Lévy–Hinčin representation. If μ is absolutely continuous (with respect to the Lebesgue measure) with an almost everywhere continuous Radon–Nikodym derivative, then p belongs to I_0^n if and only if μ is concentrated on an open set A satisfying

$$A^* \cap \bigcup_{k=2}^{\infty} (k)A = \phi \tag{7.2.1}$$

where A^* is the convex hull of A.

For the proof, we will use the following.

Lemma 7.2.1 Let $\hat{p}$ be the characteristic function defined by

$$\log \hat{p}(t) = \int (e^{i(t, u)} - 1) f(u) \lambda(du)$$

where

$$f(u) = \begin{cases} \rho & \text{if } \quad u \in A_j = \{x \in R^n : \|x - a_j\| < r\} \qquad (j = 1, \ldots, l), \\ \rho & \text{if } \quad u \in B_k = \{x \in R^n : \|x - b_k\| < r\} \qquad (k = 1, \ldots, m), \\ 0 & \text{otherwise} \end{cases}$$

($m \geq 2$, $\rho > 0$, $r > 0$, $a_j \in R^n$, $b_k \in R^n$). If there exist some positive integers λ_j satisfying

$$\left(\sum_{j=1}^{l} \lambda_j a_j\right) = \left(\sum_{j=1}^{l} \lambda_j\right)\left(\sum_{k=1}^{m} b_k\right), \tag{7.2.2}$$

then p does not belong to I_0^n.

Proof From Corollary 4.3.7, we deduce easily that we can suppose without any loss of generality that the a_j and b_k are all distinct and that

$$r < \frac{m}{m+1} \inf\left(\inf_{1 \leq j \leq l} \left\| a_j - \sum_{k=1}^{m} b_k \right\|, \inf_{1 \leq k \leq m} \left\| b_k - \sum_{j=1}^{m} b_j \right\| \right).$$

Under this condition, $C = \{x \in R^n : \|x - \sum_{k=1}^{m} b_k\| < r/m\}$ is disjoint from A_j $(j = 1, \ldots, l)$ and B_k $(k = 1, \ldots, m)$.

For $\varepsilon > 0$, we define now g_ε by

$$g_\varepsilon(x) = \begin{cases} -\varepsilon & \text{if } \quad x \in C, \\ f(x) & \text{if } \quad x \notin C, \end{cases}$$

and f_s and $g_{s,\varepsilon}$ ($s \in N$) by

$$\begin{aligned} f_1 &= f, \qquad f_s(x) = \int f_{s-1}(x-y) f(y)\lambda(dy), \\ g_{1,\varepsilon} &= g_\varepsilon, \qquad g_{s,\varepsilon}(x) = \int g_{s-1,\varepsilon}(x-y) g_\varepsilon(y)\lambda(dy). \end{aligned} \tag{7.2.3}$$

We have then

$$\lim_{\varepsilon \to 0} \sup_x (f_s(x) - g_{s,\varepsilon}(x)) = 0 \tag{7.2.4}$$

($s = 1, 2, \ldots$).

If we define L_s, M_s, and N_s by

$$L_s = \bigcup ((\alpha_1)A_1 + \cdots + (\alpha_l)A_l + (\beta_1)B_1 + \cdots + (\beta_m)B_m + (\gamma)C)$$

where the union is taken on all the positive integers γ and all the nonnegative integers α_j and β_k satisfying

$$\sum_{j=1}^{l} \alpha_j + \sum_{k=1}^{m} \beta_k + \gamma = s;$$

$$M_s = \bigcup ((\alpha_1)A_1 + \cdots + (\alpha_l)A_l + (\beta_1)B_1 + \cdots + (\beta_m)B_m)$$

where the union is taken on all the nonnegative integers α_j and β_k satisfying

$$\sum_{j=1}^{l} \alpha_j + \sum_{k=1}^{m} \beta_k = s;$$

$$N_s = M_s \backslash \{x \in R^n : \|y - x\| < r/m \text{ for some } y \in \partial M_s\}$$

where ∂M_s is the boundary of M_s, then we have

$$g_{s,\varepsilon}(x) = f_s(x) \qquad \text{if} \quad x \notin L_s \tag{7.2.5}$$

and

$$\inf_{x \in N_s} f_s(x) > 0. \tag{7.2.6}$$

From the definition and (7.2.2), we have also

$$Q_s = \bigcup_{k=1}^{s} L_k \subset \bigcup_{k=1}^{ms} N_k. \tag{7.2.7}$$

We prove now the existence of some s_0 such that $L_s \subset N_s$ if $s \geq s_0$. Indeed if A is the smallest convex set containing the a_j and b_k, there exists in A a set $\{c_\kappa\}_{\kappa=1}^M$ where c_κ is of the kind

$$\left(\sum_{j=1}^{l} \mu_j + \sum_{k=1}^{m} \nu_k\right) c_\kappa = \sum_{j=1}^{l} \mu_j a_j + \sum_{k=1}^{m} \nu_k b_k$$

($\mu_j \in N$, $\nu_k \in N$) and satisfies

$$\sup_{\kappa} \inf_{\kappa' \neq \kappa} \|c_\kappa - c_{\kappa'}\| < r/m.$$

If s_0 is the smallest common multiple of the $\sum_{j=1}^{l} \mu_j + \sum_{k=1}^{m} \nu_k$, the union of the balls with center sc and radius sr contains L_s and is contained in N_s for any $s > s_0$.

From (7.2.4), (7.2.6), and (7.2.7), we obtain the existence of an $\varepsilon_0 > 0$ such that

$$\inf_{x \in L_s} g_{s,\varepsilon}(x) \geq 0$$

($s = s_0, s_0 + 1, \ldots, 2s_0 - 1$) and

$$\inf_{x \in Q_{s_0 - 1}} \sum_{k=1}^{m(s_0 - 1)} (1/k!) g_{k,\varepsilon}(x) \geq 0$$

if $\varepsilon < \varepsilon_0$. From (7.2.5), we deduce that

$$g_{s,\varepsilon}(x) \geq 0 \tag{7.2.8}$$

($s = s_0, s_0 + 1, \ldots, 2s_0 - 1$) and

$$\sum_{k=1}^{m(s_0 - 1)} (1/k!) g_{k,\varepsilon}(x) \geq 0$$

if $\varepsilon < \varepsilon_0$ for any $x \in R^n$. From (7.2.3) we have

$$g_{s+t,\varepsilon}(x) = \int g_{s,\varepsilon}(x - y) g_{t,\varepsilon}(y) \lambda(dy)$$

so that, from (7.2.8), $g_{s,\varepsilon} \geq 0$ for any $s \geq s_0$ and

$$\sum_{k=1}^{\infty} (1/k!) g_{k,\varepsilon} \geq 0$$

if $\varepsilon < \varepsilon_0$.

If we consider the absolutely continuous signed measure ν_ε having g_ε for Radon–Nikodym derivative, then $\exp \nu_\varepsilon - \delta_0$ is absolutely continuous and has for Radon–Nikodym derivative $\sum_{k=1}^{\infty} (1/k!) g_{k,\varepsilon}$ and therefore $\exp \nu_\varepsilon$ is a measure if $\varepsilon < \varepsilon_0$. From Theorem 4.3.4, there exists a characteristic function $\hat{q}$ admitting a $(0, 0, \nu_\varepsilon)$-De Finetti representation. Since $\nu_\varepsilon(C) < 0$, it follows from Corollary 4.3.8 that q is not infinitely divisible and since $g_\varepsilon \leq f$, Corollary 4.3.7 implies that q divides p and the lemma is proved.

Proof of Theorem 7.2.1 If μ is not concentrated on an open set A satisfying (7.2.1), there exist some points $a_1, \ldots, a_l$ and $b_1, \ldots, b_m$ belonging to the support of μ and satisfying (7.2.2). Since μ is absolutely continuous, there exists some $r > 0$ such that

$$\frac{\|u\|^2}{1 + \|u\|^2} \frac{d\mu}{d\lambda}(u) \geq \rho$$

for almost every $u \in \{x \in R^n : \|x - a_j\| < r\}$ $(j = 1, \ldots, l)$ and almost every $u \in \{x \in R^n : \|x - b_k\| < r\}$ $(k = 1, \ldots, m)$. It follows then from Corollary 4.3.7 and Theorem 4.3.1 that the probability of the Lemma 7.2.1 divides p. Therefore p does not belong to I_0^n and the necessity of the condition is proved. The sufficiency follows immediately from Theorem 7.1.4.

For the case $n = 1$, from Theorem 7.2.1, we easily deduce

Theorem 7.2.2 Let $\hat{p}$ be a univariate characteristic function admitting an $(\alpha, 0, \mu)$-Lévy–Hinčin representation. If μ is absolutely continuous (with respect to the Lebesgue measure) with an almost everywhere continuous Radon–Nikodym derivative, then p belongs to I_0^1 if and only if there exists some $m > 0$ such that μ is concentrated on $]m, 2m[$ or $]-2m, -m[$.

Let $A \subset R^2$ be the set defined at the end of Section 7.1 and μ be the uniform measure on A $[\mu(B) = \lambda(B \cap A)$ for any $B \in \mathscr{B}_2]$. If we consider the characteristic function $\hat{p}$ admitting the $(0, 0, \mu)$-De Finetti representation, we deduce easily from Theorem 7.2.1 and 7.2.2 that p belongs to I_0^2, but $\mathrm{pr}_\theta\, p$ does not belong to I_0^2 for any $\theta \in R^2$. This example can be extended to n-variate probabilities with $n > 2$.

7.3 ISOMORPHISM METHOD

If $A \in \mathscr{B}_n$, we denote by $\mathscr{M}(A)$ [resp. $\mathscr{P}(A)$] the set of all the finite signed measures (resp. probabilities) concentrated on A. If $A \in F_\sigma^n$ and if q is a probability dividing $p \in \mathscr{P}(A)$, Corollary 2.5.3 implies the existence of some $\alpha \in R^n$ such that $q * \delta_\alpha$ belongs to $\mathscr{P}(A)$ (we say that $\mathscr{P}(A)$ is "factor-closed").

Now let $A \in F_\sigma^n$ and M be one of the sets N, Z, Q^+, or Q. If μ and ν belong to $\mathscr{M}((M)A)$ [resp. $\mathscr{P}((M)A)$], then Theorem 2.5.1 implies that $\mu * \nu$ belongs to $\mathscr{M}((M)A)$ [rep. $\mathscr{P}((M)A)$]. In other words, $\mathscr{M}((M)A)$ is a semigroup for the convolution and $\mathscr{P}((M)A)$ is a subsemigroup of $\mathscr{M}((M)A)$.

If $A \in F_\sigma^n$ and $A' \in F_\sigma^{n'}$ are two equipotent independent sets and if ϕ is a bijection of A on A', we can extend ϕ to a bijection of $M(A)$ on $(M)A'$ (which we denote again by ϕ) by the formula

$$\phi\left(\sum_{j=1}^{m} h_j a_j\right) = \sum_{j=1}^{m} h_j \phi(a_j) \tag{7.3.1}$$

for any $m \in N$, $h_j \in M$, $a_j \in A$. We say that ϕ is *bimeasurable* if ϕ transforms a Borel subset of $(M)A$ on a Borel subset of $(M)A'$ and ϕ^{-1} transforms a Borel subset of $(M)A'$ on a Borel subset of $(M)A$. If A and A' are countable, then any bijection of $(M)A$ on $(M)A'$ is bimeasurable. If A and A' are not enumerable, there exists a bijection of $(N)A$ on $(N)A'$ which is bimeasurable (for a proof, see Appendix A.3).

The basis of the isomorphism method is

Theorem 7.3.1 Let $A \in F_\sigma^n$ and $A' \in F_\sigma^{n'}$ be two equipotent independent Borel sets and ϕ be a bimeasurable bijection of $(M)A$ on $(M)A'$ ($M = N, Z, Q^+$, or Q). Then the mapping Φ defined by

$$\Phi(\mu)(B') = \mu(\phi^{-1}(B' \cap (M)A')) \tag{7.3.2}$$

$[\mu \in \mathscr{M}((M)A),\ B' \in \mathscr{B}_{n'}]$ is an isomorphism of the semigroup $\mathscr{M}((M)A)$ on the semigroup $\mathscr{M}((M)A')$ and $\Phi(\mathscr{P}((M)A)) = \mathscr{P}((M)A')$.

Proof If $\mu \in \mathscr{M}((M)A)$, it is clear that $\Phi(\mu)$ is well defined by (7.3.2) and belongs to $\mathscr{M}((M)A')$ and that Φ is bijective, the mapping Φ^{-1} being defined by

$$\Phi^{-1}(\mu')(B) = \mu'(\phi(B \cap (M)A))$$

$[\mu' \in \mathscr{M}((M)A'),\ B \in \mathscr{B}_n]$. Moreover,

$$\Phi(\delta_a) = \delta_{\phi(a)} \tag{7.3.3}$$

if $a \in (M)A$. If $B' \in \mathscr{B}_{n'}$ and $\mu_j \in \mathscr{M}((M)A)$ $(j = 1, 2)$, then

$$\begin{aligned}
\Phi(\mu_1 * \mu_2)(B') &= (\mu_1 * \mu_2)(\phi^{-1}(B' \cap (M)A')) \\
&= \int \mu_1[\phi^{-1}(B' \cap (M)A') - x]\mu_2(dx) \\
&= \int_{(M)A} \mu_1[\phi^{-1}(B' \cap (M)A') - x]\mu_2(dx) \\
&= \int_{(M)A'} \mu_1[\phi^{-1}(B' \cap (M)A') - \phi^{-1}(y)]\Phi(\mu_2)(dy) \\
&= \int_{(M)A'} \mu_1[\phi^{-1}(B' \cap (M)A' - y)]\Phi(\mu_2)(dy) \\
&= \int \Phi(\mu_1)((B' - y) \cap (M)A')\Phi(\mu_2)(dy),
\end{aligned}$$

so that

$$\Phi(\mu_1 * \mu_2) = \Phi(\mu_1) * \Phi(\mu_2)$$

and Φ is an isomorphism of $\mathscr{M}((M)A)$ on $\mathscr{M}((M)A')$. Now, since $\Phi(\mu)$ is a measure when μ is a measure and since

$$\|\Phi(\mu)\| = \|\mu\|,$$

$\Phi(\mu)$ is a probability when μ is a probability and the theorem is proved.

We call the application Φ defined by (7.3.2) the *isomorphism induced on* $\mathscr{M}((M)A)$ by the bijection ϕ.

Corollary 7.3.1 In the conditions of Theorem 7.3.1, $p \in \mathscr{P}((M)A)$ has no indecomposable factor if and only if $\Phi(p)$ has no indecomposable factor.

Proof If $p \in \mathscr{P}((M)A)$ and q divides p, then, from Corollary 2.5.3, there exists some $\alpha \in R^n$ such that $q * \delta_\alpha$ belongs to $\mathscr{P}((M)A)$. Therefore, we can define $\Phi(q * \delta_\alpha) = q'$ and, since Φ is an isomorphism, q' divides $\Phi(p)$. In other words, if $p \in \mathscr{P}((M)A)$, Φ transforms the set of all the factors of p belonging to $\mathscr{P}((M)A)$ (and we have in this set all the factors of p up to a degenerate factor) onto the set of all the factors of $\Phi(p)$ belonging to $\mathscr{P}((M)A')$ (and we have in this set all the factors of $\Phi(p)$ up to a degenerate factor). Therefore $p \in \mathscr{P}((M)A)$ is infinitely divisible if and only if $\Phi(p)$ is infinitely divisible and (7.3.3) implies that $p \in \mathscr{P}((M)A)$ is indecomposable if and only if $\Phi(p)$ is indecomposable. Therefore $p \in \mathscr{P}((M)A)$ has no indecomposable factor if and only if $\Phi(p)$ has no indecomposable factor.

Corollary 7.3.2 In the conditions of Theorem 7.3.1, the characteristic function $\hat{p}$ of a probability $p \in \mathscr{P}((M)A)$ admits an $(\alpha, 0, \mu)$-De Finetti representation with $\alpha \in (M)A$ and $\mu \in \mathscr{M}((M)A)$ if and only if $\Phi(p)$ admits a $(\phi(\alpha), 0, \Phi(\mu))$-De Finetti representation.

Proof Since Φ is an isomorphism, we have

$$\Phi(\exp \mu) = \exp(\Phi(\mu))$$

and this with (7.3.3) implies easily the corollary.

As application, we will prove the n-variate extensions of Theorems 4.3.9 and 4.3.10.

Theorem 7.3.2 An n-variate characteristic function $\hat{p}$ admits a $(0, 0, \mu)$-De Finetti representation with a purely atomic symmetric finite signed measure μ if and only if the following conditions are satisfied:

(a) p is a purely atomic symmetric probability;
(b) $\inf_{t \in R^n} \hat{p}(t) > 0$.

Moreover, if p is concentrated on A, then μ is concentrated on $(Z)A$.

Proof The necessity of conditions (a) and (b) is clear. Now let p be an n-variate probability satisfying conditions (a) and (b). Then p is concentrated on a countable set A. Let B be a countable independent set such that $A \subset (Q)B$. Since B is countable, the set $\Theta = \{\theta \in R^n : \mathrm{pr}_\theta B$ is not an independent set$\}$ is countable. If $\theta \notin \Theta$, then pr_θ is a bimeasurable bijection between $(Q)B$ and $(Q)\,\mathrm{pr}_\theta B$. Now $\mathrm{pr}_\theta p$ is a purely atomic symmetric probability satisfying

$$\inf_{t \in R^n} \mathrm{pr}_\theta \hat{p}(t) > 0.$$

From Theorem 4.3.9, $\mathrm{pr}_\theta p$ admits a $(0, 0, \mu')$-De Finetti representation with $\mu' \in \mathcal{M}_n$ concentrated on $(Z)\,\mathrm{pr}_\theta A$ and, from Corollary 7.3.2, $\hat{p}$ admits a $(0, 0, \mu)$-De Finetti representation with $\mu \in \mathcal{M}_n$ concentrated on $(Z)A$.

Theorem 7.3.3 An n-variate characteristic function $\hat{p}$ admits an $(\alpha, 0, \mu)$-De Finetti representation with a purely atomic finite signed measure μ satisfying

$$\int e^{-(r,u)} \,|\mu|\,(du) < +\infty$$

for $\|r\| < \rho$ $(\rho > 0)$ if and only if the following conditions are satisfied:

(a) p is a purely atomic probability;
(b) $\hat{p}$ is an analytic characteristic function regular and without zeros in the tube $T = \{t \in C^n : \|\mathrm{Im}\, t\| < \rho\}$.

Moreover, if p is concentrated on A, μ is concentrated on $(Z)A$.

Proof The case $n = 1$ is Theorem 4.3.10. The necessity can be proved exactly as in the case $n = 1$. For the sufficiency, we proceed as for the preceding theorem. Let B be a countable independent set such that $A \subset (Q)B$. Since B is countable, $\Theta = \{\theta \in R^n : \mathrm{pr}_\theta B$ is not independent$\}$ is countable. If $\theta \notin \Theta$, pr_θ is a bimeasurable bijection of $(Q)B$ on $(Q)\,\mathrm{pr}_\theta B$ and $\mathrm{pr}_\theta p$ admits an $(\alpha', 0, \mu')$-De Finetti representation with $\mu' \in \mathcal{M}_n$ concentrated on $(Z)\,\mathrm{pr}_\theta A$ and satisfying

$$\int e^{-(r,u)} \,|\mu'|\,(du) < +\infty \tag{7.3.4}$$

for $\|r\| < \rho$. From Corollary 7.3.2, $\hat{p}$ admits an $(\alpha, 0, \mu)$-De Finetti representation with $\mu \in \mathcal{M}_n$ concentrated on $(Z)A$. We have $\mu' = \mathrm{pr}_\theta\, \mu$ and since pr_θ is a bijection, we have

$$|\mu'| = |\mathrm{pr}_\theta\, \mu| = \mathrm{pr}_\theta\, |\mu|,$$

so that (7.3.4) implies

$$\int e^{-(r,u)}\, \mathrm{pr}_\theta\, |\mu|\,(du) < +\infty$$

for any $\theta \notin \Theta$ and any r, $\|r\| < \rho$. Since Θ^c is dense in R^n and $|\mu|$ is up to a multiplicative factor a probability, we deduce easily from Theorem 3.3.2 that

$$\int e^{-(r,u)}\,|\mu|\,(du) < +\infty$$

for any r, $\|r\| < \rho$.

7.4 INDEPENDENT SETS

We introduce now the following definitions.

A set $A \subset R^n$ is a *generalized independent set* of type τ if

$$A = \bigcup_{j=1}^{\infty} \bigcup_{k=r_j}^{s_j} (k)B_j$$

where $B_j \in F_\sigma^n$ ($j = 1, 2, \ldots$), $B = \bigcup_{j=1}^{\infty} B_j$ is an independent set, and r_j and s_j are integers satisfying $0 < r_j \leq s_j < \tau r_j$ ($j = 1, 2, \ldots$).

If $A \subset (Q)B$ where B is an independent set, B is a *basis* of A.

The sets $A_j \subset (Q)B_j$ ($j = 1, 2, \ldots$) have *independent basis* if $\bigcup_{j=1}^{\infty} B_j$ is an independent set.

We have then

Theorem 7.4.1 Let $\hat{p}$ be a characteristic function admitting an $(\alpha, 0, \mu)$-De Finetti representation with a signed measure μ concentrated on a generalized independent set of type τ

$$A = \bigcup_{j=1}^{\infty} \bigcup_{k=r_j}^{s_j} (k)B_j\,.$$

If q divides p, then $\hat{q}$ admits a $(\beta, 0, \nu)$-De Finetti representation with a finite signed measure ν concentrated on A and nonnegative on

$$T = \bigcup_{j=1}^{\infty} \bigcup_{k=r_j}^{2r_j-1} (k)B_j\,.$$

Proof We choose an independent set B' of real numbers which is equipotent to $B = \bigcup_{j=1}^{\infty} B_j$ and such that

$$\phi(B_j) \subset \,]1/r_j, 2/(2r_j - 1)[\qquad (j = 1, 2, \ldots),$$

ϕ meaning a bimeasurable bijection of $(N)B$ on $(N)B'$. If Φ is the isomorphism induced by ϕ on $\mathcal{M}((N)B)$, from Corollary 7.3.2, the characteristic function of $p' = \Phi(p * \delta_{-\alpha})$ admits a $(0, 0, \Phi(\mu))$-De Finetti representation. Since $\Phi(\mu)$ is concentrated on $]1, 2\tau[$, it follows from Theorem 7.1.2 that the characteristic function of any factor q' of p' admits a $(\beta', 0, \nu')$-De

Finetti representation with a finite signed measure ν' concentrated on $C(\Phi(\mu)) \cap (N)B'$ and nonnegative on $]1, 2[$. When $q' \in \mathscr{P}((N)B')$, $\beta' \in (N)B'$ and we deduce from Corollary 7.3.2 that the characteristic function of $q = \Phi^{-1}(q')$ admits a $(\beta, 0, \nu)$-De Finetti representation with

$$\beta = \phi^{-1}(\beta'), \qquad \nu = \Phi^{-1}(\nu').$$

From the properties of ν', we deduce easily that ν is nonnegative on $\phi^{-1}(]1, 2[)$ and since $\phi(T) \subset]1, 2[$, ν is nonnegative on T. Since any factor of p is of the kind $\Phi^{-1}(q') * \delta_\gamma$ where $q' \in \mathscr{P}((N)B')$ divides p' and $\gamma \in R^n$, it remains to prove that ν is concentrated on A, but this follows easily from the two following lemmas.

Lemma 7.4.1 Let $\hat{p}$ be a characteristic function admitting an $(\alpha, 0, \mu)$-De Finetti representation with a finite signed measure μ concentrated on $A = \bigcup_{k=r}^{s} (k)B$ where B is an independent set and r and s are two positive integers. If q divides p, then the Poisson measure of $\hat{q}$ is concentrated on A.

Proof We choose an independent set $B' \subset]1, 1 + (1/s)[$ which is equipotent to B and let ϕ be a bimeasurable bijection of $(N)B$ on $(N)B'$. Then ϕ induces an isomorphism Φ on $\mathscr{M}((N)B)$, and $\Phi(\mu)$ is concentrated on $\phi(A) \subset]r, s + 1[$. If $q \in \mathscr{P}((N)B)$ divides p, then $\Phi(q)$ divides $\Phi(p * \delta_{-\alpha})$. It follows then from Theorem 7.1.2 that $\Phi(q)$ has a $(\beta', 0, \nu')$-De Finetti representation with a signed measure ν' concentrated on $]r, s + 1[$. From Corollary 7.3.2, q has a $(\beta, 0, \nu)$-De Finetti representation with a signed measure ν concentrated on $\phi^{-1}(]r, s + 1[) = A$.

Lemma 7.4.2 Let $\hat{p}$ be a characteristic function admitting an $(\alpha, 0, \mu)$-De Finetti representation with a signed measure μ concentrated on a generalized independent set A of type τ.

(a) If $A = \bigcup_{j=1}^{\infty} A_j$ where A_j are sets with independent basis, then

$$\hat{p}(t) = e^{i(\alpha, t)} \prod_{j=1}^{\infty} \hat{p}_j(t)$$

where $\hat{p}_j$ is a characteristic function admitting a $(0, 0, \mu_j)$-De Finetti representation with a signed measure μ_j concentrated on A_j ($j = 1, 2, \ldots$).

(b) If q divides p, then

$$\hat{q}(t) = e^{i(\beta, t)} \prod_{j=1}^{\infty} \hat{q}_j(t)$$

where $\beta \in R^n$ and q_j is a probability dividing p_j ($j = 1, 2, \ldots$).

Proof Let B_j be a basis of A_j and $B = \bigcup_{j=1}^{\infty} B_j$. If q divides p, it follows from above that $\hat{q}$ admits a $(\beta, 0, \nu)$-De Finetti representation with a signed measure ν concentrated on $(N)B$.

If m is a positive integer, we consider now an independent set $B'_m \in F^2_\sigma$ which is equipotent to B and has the following properties [ϕ_m means a bimeasurable bijection of $(N)B$ on $N(B'_m)$]:

$$\phi_m(B_m) \subset \{(x, y) \in R^2 : y = 0, 1 < x < 1 + s_m^{-1}\},$$

$$\phi_m(B_j) \subset \{(x, y) \in R^2 : x = 0, 1/r_j < y < 2/(2r_j - 1)\} \qquad \text{if } j \neq m.$$

If Φ_m is the isomorphism induced by ϕ_m on $\mathscr{M}((N)B)$, we put

$$p' = \Phi_m(p * \delta_{-\alpha}), \qquad q' = \Phi_m(q * \delta_{-\beta})$$

and obtain from Theorem 2.9.2 that

$$p' = p'_{m,1} \otimes p'_{m,2}$$

where $p'_{m,1}$ and $p'_{m,2}$ admit De Finetti representations with signed measures concentrated on $T_1 = \{(x, y) \in R^2 : y = 0, \quad r_m < x < s_m + 1\}$ and $T_2 = \{(x, y) \in R^2 : x = 0, 1 < y < 2\tau\}$, respectively, and from Theorem 5.3.4 we deduce that

$$q' = q'_{m,1} \otimes q'_{m,2}$$

where $q'_{m,j}$ divides $p'_{m,j}$ $(j = 1, 2)$. From Corollary 7.3.2, we deduce that

$$p = \delta_\alpha * p_{m,1} * p_{m,2}$$

where $\hat{p}_{m,1}$ and $\hat{p}_{m,2}$ admit De Finetti representations with signed measures concentrated on A_m and $A \backslash A_m$, respectively, and

$$q = \delta_\beta * q_{m,1} * q_{m,2}$$

where $q_{m,j}$ divides $p_{m,j}$ $(j = 1, 2)$. Since m is arbitrary, this implies the lemma by induction.

From Theorem 7.4.1, we deduce immediately

Corollary 7.4.1 If the characteristic function $\hat{p}$ admits an $(\alpha, 0, \mu)$-De Finetti representation with a Poisson measure μ concentrated on a generalized independent set of type 2, then p belongs to I_0^n.

A particular case of this corollary is the important

Theorem 7.4.2 If the characteristic function $\hat{p}$ admits an $(\alpha, 0, \mu)$-De Finetti representation with a Poisson measure concentrated on an independent set, then p belongs to I_0^n.

Corollary 7.4.2 The set I_0^n of all the n-variate probabilities without indecomposable factors is dense (with respect to complete convergence) in the set I^n of all the n-variate infinitely divisible probabilities.

Proof From Theorem 4.2.4, it suffices to prove that an infinitely divisible probability p admitting a $(0, 0, \mu)$-De Finetti representation is the complete limit of a sequence of probabilities belonging to I_0^n. In the cube

$$A_{j, k_1, \ldots, k_n} = \{x = (x_1, \ldots, x_n) : k_m/j \leq x_m < (k_m + 1)/j, m = 1, \ldots, n\}$$

$(j = 1, 2, \ldots; k_m = 0, \pm 1, \pm 2, \ldots, \pm j^2)$, we choose a point $\lambda_{j, k_1, \ldots, k_n}$ so that $\Lambda_j = \{\lambda_{j, k_1, \ldots, k_n}\}$ is an independent set for each j and we consider the probability p_j admitting the $(0, 0, \mu_j)$-De Finetti representation where μ_j is the measure defined by

$$\mu_j(B) = \sum_{\lambda_{j, k_1, \ldots, k_n} \in B} \mu(A_{j, k_1, \ldots, k_n})$$

for any $B \in B_n$. Evidently $\mu_j \xrightarrow{c} \mu$ and, from Corollary 2.9.5, we deduce that $p_j \xrightarrow{c} p$. Since μ_j is concentrated on the set Λ_j, we obtain from Theorem 7.4.2 that $p_j \in I_0^n$ $(j = 1, 2, \ldots)$ and the corollary is proved.

In connection with this result which extends Theorem 7.1.3, we must note that $I^n \backslash I_0^n$ is also dense in I^n (this follows easily from Theorem 7.2.1).

Corollary 7.4.3 (a) If Γ is a convex set containing the origin, there exists a probability which belongs to $I_0^n \cap \mathscr{A}_\Gamma$, but does not belong to $\mathscr{A}_{\Gamma'}$, for any convex set Γ' greater than Γ.

(b) There exists a probability which belongs to I_0^n and does not belong to $\mathscr{A}_\Gamma$ for any convex set Γ containing the origin.

Proof In each cube

$$A_{k_1, \ldots, k_n} = \{x = (x_1, \ldots, x_n) : k_m \leq x_m < k_m + 1, m = 1, \ldots, n\}$$

$[(k_1, \ldots, k_n) \in Z^n]$, we choose a point $\lambda_{k_1, \ldots, k_n}$ so that $\Lambda = \{\lambda_{k_1, \ldots, k_n}\}$ is an independent set. If μ is a measure, we define a measure v by

$$v(B) = \sum_{\lambda_{k_1, \ldots, k_n} \in B} \mu(A_{k_1, \ldots, k_n})$$

for any $B \in B_n$. Then v is concentrated on Λ and $v/\|v\|$ belongs to $\mathscr{A}_\Gamma$ if and only if $\mu/\|\mu\|$ belongs to $\mathscr{A}_\Gamma$. If we consider the characteristic function $\hat{p}$ admitting the $(0, 0, \mu)$-De Finetti representation, then, from Theorem 7.4.2, $\hat{p}$ belongs to I_0^n and, from Theorem 5.2.5, $\hat{p}$ belongs to $\mathscr{A}_\Gamma$ if and only if $\mu/\|\mu\|$ belongs to $\mathscr{A}_\Gamma$. The corollary follows from the existence of probabilities which belong to $\mathscr{A}_\Gamma \backslash \bigcup_{\Gamma' \supset \Gamma, \Gamma' \neq \Gamma} \mathscr{A}_{\Gamma'}$ or which do not belong to $\mathscr{A}_\Gamma$ for any convex set Γ.

Corollary 7.4.4 Any closed set $A \subset R^n$ of the kind $A = B \cup C$ where B and C are disjoint, B being a perfect set and C being either empty or a countable independent set, is the Poisson spectrum of a probability $p \in I_0^n$.

Proof There exists an enumerable set B' which is dense in B and such that $B' \cup C$ is independent. Then $B' \cup C = \{m_k\}$ and if we consider the measure μ defined by

$$\mu(B) = \sum_{m_k \in B} \alpha_k$$

where $\alpha_k > 0, \sum_k \alpha_k < +\infty$, then the probability p admitting the $(0, 0, \mu)$-De Finetti representation belongs to I_0^n and has A as Poisson spectrum.

We extend now Corollary 7.4.1 to

Theorem 7.4.3 Let $\hat{p}$ be a characteristic function admitting an $(\alpha, 0, \mu)$-De Finetti representation with a Poisson measure μ concentrated on $A \in F_\sigma^n$. If the projection A_m of A on an m-dimensional subspace E of R^n is a generalized independent set of type 2, then p belongs to I_0^n.

Proof Let F be the subspace such that

$$E \oplus F = R^n.$$

If we choose a basis in F, we define a natural bijection ψ of F on R^{n-m}. If

$$A_m = \bigcup_{j=1}^{\infty} \bigcup_{k=r_j}^{s_j} (k)B_j$$

where $B_j \in F_\sigma^n$, $B = \bigcup_{j=1}^{\infty} B_j$ is an independent set, and r_j and s_j are positive integers satisfying $r_j \leq s_j < 2r_j$ $(j = 1, 2, \ldots)$, we consider an independent set $B' \in F_\sigma^1$ which is equipotent to B. If ϕ is a bimeasurable bijection of $(N)B$ on $(N)B'$, we can define an application θ of $C = (N)B \times F$ on $C' = (N)B' \times R^{n-m}$ by the formula

$$\theta(x, y) = (\phi(x), \psi(y))$$

for any $x \in (N)B$, $y \in F$. This application is a bimeasurable bijection of C on C' and with this bijection we can define an isomorphism Θ of $\mathscr{M}(C)$ on $\mathscr{M}(C')$ by a formula analogous to (7.3.2) and this isomorphism has all the properties stated in Section 7.3.

If we choose $\phi(B_j) \subset]1/r_j, 2/(2r_j - 1)[$ $(j = 1, 2, \ldots)$, then $\Theta(p)$ has a Poisson measure concentrated on the set $\{y = (y_1, \ldots, y_{n-m+1}): 1 < y_1 < 2\}$. It follows from Corollary 7.1.1 that $\Theta(p) \in I_0^{n-m+1}$, so that $p \in I_0^n$.

7.5 INDEPENDENT SETS AND PROJECTIONS

Let $A \in F_\sigma^n$ be an independent set. If $E \subset A$, then any $x \in (Q)A$ admits an unique decomposition

$$x = y + z$$

where $y \in (Q)E$ and $z \in (Q)(A \backslash E)$. We call y the *projection* of x on E and we write

$$y = P_E(x).$$

If $\mu \in \mathscr{M}((Q)A)$, we define $P_E(\mu)$ by

$$P_E(\mu)(B) = \mu(\{x \in (Q)A : P_E(x) \in B\})$$

for any $B \in B_n$. Thus $P_E(\mu)$ is the projection of μ on E. Evidently, $P_E(\mu) \in \mathscr{M}((Q)E)$ and $P_E(\mu)$ is a probability if μ is a probability.

We prove now that this projection has some of the properties of the projections on real subspaces defined in Section 2.4. For example, we have

Theorem 7.5.1 If $A \in F_\sigma^n$ is an independent set and $E \subset A$, then

$$P_E(\mu_1 * \mu_2) = P_E(\mu_1) * P_E(\mu_2).$$

Proof We consider in R^{2n} the set $A' = \{(x, y) : x \in (Q)E,\ y \in (Q)(A \backslash E)\}$. Then if ϕ is the application of R^{2n} on R^n defined by

$$\phi(x, y) = x + y,$$

ϕ is a bimeasurable bijection of $(Q)A'$ on $(Q)A$. If Φ is the isomorphism induced by ϕ on $\mathscr{M}((Q)A')$ and if $E' = \{z = (z_1, \dots, z_{2n}) \in R^{2n} : z_{n+1} = \cdots z_{2n} = 0\}$, then

$$P_E = \Phi^{-1} \circ \mathrm{pr}_{E'} \circ \Phi$$

and the theorem follows from the fact that Φ is an isomorphism and from Corollary 2:5.8.

Corollary 7.5.1 Let $A \in F_\sigma^n$ be an independent set and $p \in \mathscr{P}((Q)A)$. If $E \subset A$ and if q divides p, then $P_E(q)$ divides $P_E(p)$.

Corollary 7.5.2 Let $A \in F_\sigma^n$ be an independent set and $p \in \mathscr{P}((Q)A)$. If $E \subset A$ and if p is infinitely divisible, then $P_E(p)$ is infinitely divisible.

Lemma 7.5.1 Let $A \in F_\sigma^n$ be an independent set and $\{E_j\}$ be an increasing sequence of sets such that

$$A = \bigcup_{j=1}^{\infty} E_j .$$

If $p \in \mathscr{P}((Q)A)$, then

$$\lim_{j \to \infty} P_{E_j}(p) \stackrel{c}{=} p.$$

Proof If $B \subset (Q)A$, then

$$B = \bigcup_{j=1}^{\infty} B_j$$

where $B_j = B \cap (Q)E_j$, so that if $B \in \mathscr{B}_n$,

$$\lim_{j\to\infty} p(B_j) = p(B). \tag{7.5.1}$$

Likewise, from

$$\bigcap_{j=1}^{\infty} B'_j = \varnothing$$

where $B'_j = \{x \in A : P_{E_j}(x) \in (Q)B_j, x \notin (Q)B_j\}$, we have

$$\lim_{j\to\infty} p(B'_j) = 0. \tag{7.5.2}$$

Since

$$P_{E_j}(p)(B) = p(B_j) + p(B'_j),$$

we have from (7.5.1) and (7.5.2) that

$$\lim_{j\to\infty} P_{E_j}(p)(B) = p(B)$$

for any $B \in \mathscr{B}_n$ and the lemma is proved.

Theorem 7.5.2 Let $A \in F_\sigma^n$ be an independent set and $\{E_j\}$ be an increasing sequence of sets such that

$$A = \bigcup_{j=1}^{\infty} E_j .$$

If the projection $P_{E_j}(p)$ of a probability $p \in \mathscr{P}((Q)A)$ has no indecomposable factor ($j = 1, 2, \ldots$), then p has no indecomposable factor.

Proof Let q be a probability dividing p. From Corollary 7.5.1, $P_{E_j}(q)$ divides $P_{E_j}(p)$ and, by assumption, $P_{E_j}(q)$ is infinitely divisible. From Lemma 7.5.1, q is the weak limit of $P_{E_j}(q)$ and q is infinitely divisible, so that $p \in I_0^n$.

We now extend Theorem 7.4.3 to

Theorem 7.5.3 Let $\hat{p}$ be a univariate characteristic function admitting an $(\alpha, 0, \mu)$-De Finetti representation with a Poisson measure μ concentrated on $A \in F_\sigma^n$. If the projection A_m of A on an m-dimensional subspace E of R^n is contained in $(Q)B$ with an independent set B and if there exists some $F \subset B$ such that $P_F(A_m)$ is a generalized independent set of type 2, then p belongs to I_0^n.

Proof If we choose in $E^{\perp}$ a basis, we define a natural bijection ψ of $E^{\perp}$ on R^{n-m}. We define then a mapping θ of $(Q)B \oplus E^{\perp} = \mathrm{C} \subset R^n$ in R^{n+m} by

$$\theta(x) = (P_F(x), P_{B\setminus F}(x), \psi(x)).$$

Then θ is a bimeasurable bijection of C on $C' = \theta(C)$ and, with θ, we can define an isomorphism Θ of $\mathscr{M}(C)$ on $\mathscr{M}(C')$ by a formula analogous to (7.3.2) and this isomorphism has all the properties stated in Section 7.3. Then $\Theta(p)$ admits a $(\theta(\alpha), 0, \Theta(\mu))$-De Finetti representation. Since $\Theta(\mu)$ is concentrated on $\theta(A)$ and since $\mathrm{pr}_G\, \theta(A)$ is a generalized independent set of type 2 (G being the subspace generated by the m first vectors of the canonical basis in R^{n+m}), it follows from Theorem 7.4.3 that $\Theta(p)$ belongs to I_0^{n+m} and therefore p belongs to I_0^n.

NOTES

The results of this chapter are centered around three axes of searches. The first one is essentially concerned with the results of Section 7.1. Its origin is the work of Raĭkov (1938a) who proved Corollary 7.1.1 when $n = 1$ and the Poisson spectrum is finite. This result was extended to the case $n = 1$ and arbitrary Poisson spectrum by Ostrovskiĭ (1965a) and to the case $n > 1$ and bounded Poisson spectrum by Ostrovskiĭ (1966b) and Cuppens (1967). Theorem 7.1.2 is due to Cuppens (1969f) and Theorem 7.1.4 to Livsič (1972). The example following Theorem 7.1.4 is taken from Linnik and Ostrovskiĭ (1972). Theorems 7.1.5–7.1.7 are new, but the particular case of Theorem 7.1.6 when $n = 1$ is due to Ostrovskiĭ (1966b).

A second kind of result are those on absolutely continuous Poisson measures given in Section 7.2. The first result in this area is due to Cramér (1949) who proved the following result: if μ is a univariate absolutely continuous measure such that $d\mu(x)/d\lambda \geq C > 0$ for almost every $x \in [0, a]$ $(a > 0)$, the probability $\exp \mu$ has an indecomposable factor. This result was extended to the segment $[a, b]$ $(b > 2a)$ by Shimizu (1964). Theorem 7.2.2 is due to Cuppens (1969d,e). The case $n > 1$ was treated by Cuppens (1969f), but the result given in this note was false and the correct version (Theorem 7.2.1) is due to Livsič (1972) and Ostrovskiĭ (1972), the proof given here being the one of Cuppens (1969f). Shimizu (1964) has also given some related results using the notion of similar points of a measure. These results were extended by Cuppens (1971), but they were not complete and we do not give them here.

The third problem is the study of independent Poisson spectra and the main result is Theorem 7.4.2. In this direction, the first result is due again to

Raĭkov (1938a) who proved this result for the case $n = 1$ with positive finite Poisson spectrum. Then Lévy (1938b) proved the result for the case $n = 1$ with finite Poisson spectrum and Ostrovskiĭ (1965a) proved the case $n = 1$ with positive bounded Poisson spectrum. Other various extensions are due to Ostrovskiĭ (1966b, 1970), Cuppens (1968, 1969b,g, 1970), and Čistjakov (1971); but the general version, using isomorphism method, is due to Cuppens (1972a). The problem of bimeasurability of the bijection which is necessary for the validity of the method is stated and solved by Linnik and Ostrovskiĭ (1972). The other results obtained by isomorphism method are due to Cuppens (1972a,b,c, 1973). Theorems 7.3.2 and 7.3.3 are due to Livsič and Ostrovskiĭ (1971), but the proof given here is due to Rousseau (1974). Corollary 7.4.2 is due to Livsič and Ostrovskiĭ (1971). The belonging to I_0^n of nonanalytic characteristic functions is due to Čistjakov (1971) and Cuppens (1972a). Corollary 7.4.4 is due to Ostrovskiĭ (1970). The results given in Section 7.5 are new. Theorem 7.5.2 takes inspiration from an analogous result proved in another context by Rousseau (1973).

Chapter 8

Infinitely Divisible Probabilities with Countable Poisson Spectrum

8.1 THE GENERAL CASE

Theorem 8.1.1 Let $\hat{p}$ be an n-variate characteristic function admitting an $(\alpha, 0, \mu)$-De Finetti representation with $\alpha \in R^n$ and $\mu \in \mathscr{M}_n$ concentrated on a countable set A and satisfying

$$\int e^{(y,\, u)} \,|\, \mu \,|\, (du) < +\infty$$

for any y satisfying $\|y\| < \rho$. If q divides p, then $\hat{q}$ admits a $(\beta, 0, \nu)$-De Finetti representation with $\beta \in R^n$ and $\nu \in \mathscr{M}_n$ concentrated on $(Z)A$ and satisfying

$$\int e^{(y,\, u)} \,|\, \nu \,|\, (du) < +\infty \tag{8.1.1}$$

for any y satisfying $\|y\| < \rho$.

Proof From Theorem 7.3.3, p is a purely atomic probability concentrated on $(N)A + \alpha$ and $\hat{p}$ is an analytic characteristic function regular in $T = \{t \in R^n : \|\text{Im}\ t\| < \rho\}$. If q divides p, from Corollary 2.5.3, q is a purely atomic probability concentrated on $(N)A + \beta$ for some $\beta \in R^n$ and, from Theorem 5.3.1, $\hat{q}$ is an analytic characteristic function regular in T. It follows then from Theorem 7.3.3 that $\hat{q}$ admits a $(\beta, 0, \nu)$-De Finetti representation with a signed measure ν concentrated on $(Z)A$ and satisfying (8.1.1) for any y satisfying $\|y\| < \rho$.

Theorem 8.1.2 Let $\hat{p}$ be an n-variate characteristic function admitting an $(\alpha, 0, \mu)$-De Finetti representation with $\alpha \in R^n$ and $\mu \in \mathcal{M}_n$ satisfying the following conditions:

(a) μ is concentrated on a Linnik set Λ;
(b) for some $K > 0$

$$|\mu|(\{x : \|x\| > u\}) = O(\exp(-Ku^2))$$

when $u \to +\infty$.

If q divides p, then $\hat{q}$ admits a $(\beta, 0, \nu)$-De Finetti representation with $\beta \in R^n$ and $\nu \in \mathcal{M}_n$ concentrated on Λ and satisfies (8.1.1) for any $y \in R^n$.

Proof We suppose first that Λ is a rational Linnik set constructed on the canonical basis of R^n. From Theorem 8.1.1, it follows that $\hat{q}$ admits a $(\beta, 0, \nu)$-De Finetti representation with $\beta \in R^n$ and $\nu \in \mathcal{M}_n$ concentrated on Q^n and satisfies (8.1.1) for any $y \in R^n$. In other words

$$\log \hat{q}(t) = i(\beta, t) + \sum_{\lambda \in Q^n} m_\lambda (e^{i(\lambda, t)} - 1) \tag{8.1.2}$$

for any $t \in C^n$.

Now, let p_y, q_y, μ_y, and ν_y be defined by

$$\hat{p}_y(t) = \hat{p}(t + iy)/\hat{p}(iy), \qquad \hat{q}_y(t) = \hat{q}(t + iy)/\hat{q}(iy),$$
$$\hat{\mu}_y(t) = \hat{\mu}(t + iy), \qquad \hat{\nu}_y(t) = \hat{\nu}(t + iy)$$

for any $y \in R^n$. From Corollary 5.2.2, we deduce that p_y and q_y are probabilities and that q_y divides p_y for any $y \in R^n$. This implies with Corollary 2.5.8 that $\mathrm{pr}_{e_j} q_y$ divides $\mathrm{pr}_{e_j} p_y$ for any $y \in R^n$ and $j = 1, \ldots, n$. But $\mathrm{pr}_{e_j} p_y$ and $\mathrm{pr}_{e_j} q_y$ admit, respectively, a $(\mathrm{pr}_{e_j} \alpha, 0, \mathrm{pr}_{e_j} \mu_y)$- and a $(\mathrm{pr}_{e_j} \beta, 0, \mathrm{pr}_{e_j} \nu_y)$-De Finetti representation. Moreover $\mathrm{pr}_{e_j} \mu_y$ is concentrated on Λ_j and satisfies

$$|\mathrm{pr}_{e_j} \mu_y|(\{x : \|x\| > u\}) = O(\exp(-Ku^2))$$

when $u \to +\infty$. Therefore $\mathrm{pr}_{e_j} p_y$ satisfies for any $y \in R^n$ the assumptions of Corollary 6.1.1 and, from this result, we deduce that $\mathrm{pr}_{e_j} \nu_y$ is concentrated on Λ_j for any $y \in R^n$ and $j = 1, \ldots, n$.

From (8.1.2), we deduce that

$$\mathrm{pr}_{e_y} \nu_y(\{\alpha e_j\}) = \sum_{\substack{\{\lambda = (\lambda_1, \ldots, \lambda_n) \\ \in Q^n :\, \lambda_j = \alpha\}}} m_\lambda e^{-(\lambda, y)}$$

if $\alpha \in Q$. Therefore

$$\sum_{\substack{\{\lambda=(\lambda_1, \ldots, \lambda_n) \\ \in Q^n:\ \lambda_j=\alpha\}}} m_\lambda e^{-(\lambda, y)} = 0 \tag{8.1.3}$$

for any $y \in R^n$ if $\alpha \notin \Lambda_j$. But, from (8.1.1), we deduce easily that the first side of (8.1.3) is an entire function of y, so that (8.1.3) holds for any $y \in C^n$. From the uniqueness of Fourier series of almost periodic functions (cf. Appendix A.2), we obtain that $m_\lambda = 0$ if $\text{pr}_{e_j} \lambda \notin \Lambda_j$. Since j is arbitrary, this implies that v is concentrated on Λ and the theorem is proved in this case.

In the general case, we can define $\theta_1, \ldots, \theta_n$ and m $(0 \le m \le n)$ so that

(a) $\{\lambda_{j,k}\} \subset Q$ for $j = m + 1, \ldots, n$;
(b) there exist some k_j and k'_j such that

$$\lambda_{j,k_j} > 0, \qquad \lambda_{j,k_{j'}} < 0, \qquad \lambda_{j,k_j}\lambda_{j,k_{j'}}^{-1} \notin Q$$

$(j = 1, \ldots, m)$. Then $A = \{\lambda_{1,k_1}\theta_1, \ldots, \lambda_{m,k_m}\theta_m, \lambda_{1,k_1'}\theta_1, \ldots, \lambda_{m,k_{m'}}\theta_m, \theta_{m+1}, \ldots, \theta_n\}$ is an independent set and $\Lambda \subset (Q)A$. If we define an application ϕ of A in Q^{n+m} by

$$\begin{aligned} \phi(\lambda_{j,k_j}\theta_j) &= e_j, & j &= 1, \ldots, m, \\ \phi(\lambda_{j,k_{j'}}\theta_j) &= e_{m+j}, & j &= 1, \ldots, m, \\ \phi(\theta_j) &= e_{m+j}, & j &= m + 1, \ldots, n, \end{aligned}$$

and if we extend this application linearly to an application (denoted again by ϕ) defined on $(Q)A$, then ϕ is a bijection of $(Q)A$ on $\phi((Q)A) \subset Q^{n+m}$.

If Φ is the isomorphism induced by ϕ on $\mathcal{M}((Q)A)$, then $\Phi(p)$ admits a $(\phi(\alpha), 0, \Phi(\mu))$-De Finetti representation and $\Phi(\mu) \in \mathcal{M}_{m+n}$ is concentrated on the Linnik set $\phi(\Lambda) \subset Q^{m+n}$ and satisfies

$$|\Phi(\mu)|(\{x : \|x\| > u\}) = O(\exp(-Ku^2))$$

when $u \to +\infty$. Now, if q divides p, $\Phi(q)$ divides $\Phi(p)$ and it follows from above that $\Phi(q)$ admits a $(\beta', 0, v')$-De Finetti representation with $v' \in \mathcal{M}_{m+n}$ concentrated on $\phi(\Lambda)$ and satisfying

$$\int e^{(y', u')} |v'|(du') < +\infty \tag{8.1.4}$$

for any $y' \in R^{m+n}$. From the properties of Φ, we deduce that q admits a $(\beta, 0, v)$-De Finetti representation with $\beta = \phi^{-1}(\beta')$ and $v = \Phi^{-1}(v')$. Therefore v is concentrated on $\phi^{-1}(\phi(\Lambda)) = \Lambda$. Since $|v| = \Phi^{-1}(|v'|)$, we deduce from (8.1.4) the validity of (8.1.1) for any $y \in R^n$.

From theorems 8.1.2 and 6.4.1, we deduce immediately

Theorem 8.1.3 Let $\hat{p}$ be an n-variate characteristic function admitting an $(\alpha, 0, \mu)$-De Finetti representation. If $\mu \in \mathcal{M}_n$ satisfies the following conditions:

(a) μ is concentrated on a Linnik set Λ;
(b) each point of Λ has an issue with respect to μ;
(c) for some $K > 0$

$$|\mu|(\{x : \|x\| > u\}) = O(\exp(-Ku^2))$$

when $u \to +\infty$,

then p belongs to I_0^n.

With the method used for Corollary 6.4.1, we deduce from Theorem 8.1.3

Corollary 8.1.1 Let $\lambda_k = (\lambda_{k,1}, \ldots, \lambda_{k,n}) \in R^n$ $(k = 0, \pm 1, \pm 2, \ldots)$ be such that $\lambda_{k',j}/\lambda_{k,j}$ is either negative or an integer greater than one $(k' > k; j = 1, \ldots, n)$ and

$$l_k = (l^{(k)}_{1,0,\ldots,0}, l^{(k)}_{0,1,0,\ldots,0}, \ldots, l^{(k)}_{1,\ldots,1}) \in (R^+)^{2n-1}$$

be such that

$$\sum_{k=-\infty}^{+\infty} \left(\sum_{\varepsilon} l_\varepsilon^{(k)}\right) < +\infty \qquad \text{and} \qquad \sum_{\{\|\lambda_k\| > u\}} \left(\sum_{\varepsilon} l_\varepsilon^{(k)}\right) = O(\exp(-Ku^2))$$

when $u \to +\infty$ for some $K > 0$ $[\varepsilon = (\varepsilon_1, \ldots, \varepsilon_n)$ with $\varepsilon_j = 0$ or 1 and $\sum_\varepsilon$ indicates the summation on the $2^n - 1$ values of $\varepsilon \neq 0]$. Then the sequence of probabilities $\{q_{j,k}\}$ defined by

$$q_{j,k} = \mathfrak{p}(\lambda_{-j}, l_{-j}) * \mathfrak{p}(\lambda_{-j+1}, l_{-j+1}) * \cdots * \mathfrak{p}(\lambda_k, l_k)$$

converges completely when $j \to +\infty$ and $k \to +\infty$ to a probability p belonging to I_0^n.

8.2 LATTICE PROBABILITIES

In the case of lattice probabilities, the results of the preceding section can be refined. For example, Theorem 8.1.1 admits the following extension.

Theorem 8.2.1 Let p be an n-variate lattice probability satisfying the following conditions:

(a) $\hat{p}$ has no zeros in R^n;
(b) $D^{2(e_1+\cdots+e_n)}\hat{p}(0)$ exists.

If q divides p, then $\hat{q}$ admits a $(\beta, 0, \nu)$-De Finetti representation. Moreover, if p is concentrated on $(Z)A + \alpha$, then ν is concentrated on $(Z)A$.

Proof If q divides p, from Corollary 2.5.3, q is a lattice probability. It is clear that $\hat{q}$ has no zeros in R^n and, from Corollary 5.1.1, $D^{2(e_1+\cdots+e_n)}\hat{q}(0)$ exists. The result follows then from Theorem 4.3.8.

For the case $n = 1$, the main result of this section is

Theorem 8.2.2 Let p be a univariate probability concentrated on $\{\alpha + km\}_{k=-\infty}^{+\infty}$ with a characteristic function admitting an $(\alpha, 0, \mu)$-De Finetti representation. If μ is concentrated on a Linnik set $\Lambda \subset \{km\}_{k=-\infty}^{+\infty}$ and satisfies

$$|\mu|(\{km\}) = o(\exp(-2|k| \log |k|)) \tag{8.2.1}$$

when $|k| \to +\infty$, then p belongs to I_0^1.

Proof We can suppose without loss of generality that $m = 1$, so that $\Lambda \subset Z$. We can enumerate Λ so that $\Lambda = \{\lambda_j\}_{j=-\infty}^{+\infty}$ with $\lambda_0 = 0$ and $\lambda_j < \lambda_{j+1}$ $(j = 0, \pm 1, \pm 2, \ldots)$. From condition (c), we deduce that $\hat{p}$ is an entire function, so that

$$\log \hat{p}(t) = i\alpha t + \sum_{j=-\infty}^{+\infty} \mu(\{\lambda_j\})e^{i\lambda_j t}$$

for any $t \in C$. Letting $e^{it} = x$, from the link between the growth of a power series and its coefficients, we deduce easily that

$$|\log \hat{p}(t)| \leq \exp(o(e^{|t|/2})) \tag{8.2.2}$$

when $|t| \to +\infty$. If ρ is defined by

$$\rho(x, y) = \mathrm{Re}(\log \hat{p}(iy) - \log \hat{p}(x + iy))$$

for any $(x, y) \in R^2$, we have

$$\rho(x, y) = \sum_{j=-\infty}^{+\infty} \mu(\{\lambda_j\}) \sin^2(\tfrac{1}{2}\lambda_j x)e^{-\lambda_j y}.$$

Since Λ is a Linnik set, we have for $j \geq 0$

$$\rho(2\pi/\lambda_{j+1}, -y) = O(e^{\lambda_j y}) \tag{8.2.3a}$$

and for $j \leq 0$

$$\rho(2\pi/\lambda_{j-1}, y) = O(e^{\lambda_j y}) \tag{8.2.3b}$$

when $y \to +\infty$.

If q divides p, from Theorem 8.2.1,

$$\log \hat{q}(t) = i\beta t + \sum_{k=-\infty}^{+\infty} \nu(\{k\})e^{ikt} \tag{8.2.4}$$

for any $t \in R$. Moreover $\hat{q}$ is an entire characteristic function and from the uniqueness of the De Finetti representation and Theorem 8.1.1, we deduce that (8.2.4) holds for any $t \in C$. If we define σ by

$$\sigma(x, y) = \operatorname{Re}(\log \hat{q}(iy) - \log \hat{q}(x + iy))$$

for any $(x, y) \in R^2$, we have

$$\sigma(x, y) = \sum_{k=-\infty}^{+\infty} \nu(\{k\}) \sin^2(\tfrac{1}{2}kx)e^{-ky} \tag{8.2.5}$$

and these relations define σ as an entire function of y for any $x \in R$. From Theorem 5.2.2 and (8.2.2), we deduce easily that

$$|\sigma(x, y)| \leq \exp(o(e^{|y|/2})) \tag{8.2.6}$$

when $|y| \to +\infty$. Now, from Corollary 5.2.3, we have

$$0 \leq \sigma(x, y) \leq \rho(x, y) \tag{8.2.7}$$

for any $(x, y) \in R^2$. For $j \geq 0$, (8.2.3a) and (8.2.7) imply

$$\sigma(2\pi/\lambda_{j+1}, -y) = O(e^{\lambda_j y}) \tag{8.2.8}$$

when $y \to +\infty$. From (8.2.5), (8.2.7), and (8.2.8), using Theorem 3.6.1

$$\nu(\{k\}) \sin^2(k\pi/\lambda_{j+1}) = 0$$

for $k > \lambda_j$, so that

$$\nu(\{k\}) = 0$$

if $\lambda_j < k < \lambda_{j+1}$. Now if $j > 0$

$$\sigma(2\pi/\lambda_{j+1}, -y) = (\nu(\{\lambda_j\}) + o(1))e^{\lambda_j y}$$

when $y \to +\infty$ and this with (8.2.7) implies

$$\nu(\{\lambda_j\}) \geq 0.$$

In the same way, using (8.2.3b) instead of (8.2.3a), we have

$$\nu(\{k\}) = 0$$

if $\lambda_{j-1} < k < \lambda_j$ ($j \leq 0$) and

$$\nu(\{\lambda_j\}) \geq 0$$

if $j < 0$. Since j is arbitrary, this implies that q is infinitely divisible and p belongs to I_0^1.

From this result, we deduce immediately

Corollary 8.2.1 Let p be a univariate probability satisfying the assumptions of Theorem 8.2.2. If q divides p, then $\hat{q}$ admits a $(\beta, 0, \nu)$-De Finetti representation with a finite measure ν concentrated on Λ and satisfies for any $\lambda \in \Lambda$

$$0 \leq \nu(\{\lambda\}) \leq \mu(\{\lambda\}).$$

We prove now

Theorem 8.2.3 The univariate probability p defined by

$$\log \hat{p}(t) = \sum_{j=1}^{\infty} \exp(-2^j)(\exp(2^j it) - 1)$$

does not belong to I_0^1.

Proof We prove first some lemmas.

Lemma 8.2.1 If

$$\phi(x) = \exp\left(\sum_{j=0}^{\infty} x^{2^j}\right) = \sum_{k=0}^{\infty} a_k x^k,$$

then

$$a_{k+1}/a_k > \tfrac{2}{3}. \tag{8.2.9}$$

Proof Indeed, ϕ satisfies the equation

$$\phi(x) = e^x \phi(x^2)$$

which implies

$$\sum_{k=0}^{\infty} a_k x^k = \left(\sum_{l=0}^{\infty} x^l/l!\right)\left(\sum_{m=0}^{\infty} a_m x^{2m}\right)$$

and from this we obtain the induction formulas

$$a_{2k} = \sum_{j=0}^{k} \frac{a_{k-j}}{(2j)!}, \qquad a_{2k+1} = \sum_{j=0}^{k} \frac{a_{k-j}}{(2j+1)!}.$$

We proceed now by induction. Since $a_0 = a_1 = 1$, (8.2.9) holds for $k = 0$. We suppose now that (8.2.9) is satisfied for $k < K$ and prove (8.2.9) for $k = K$. We suppose first $K = 2l$ even. Then

$$\begin{aligned}\frac{a_{K+1}}{a_K} = \frac{a_{2l+1}}{a_{2l}} &= \left(\sum_{j=0}^{l} \frac{a_{l-j}}{(2j+1)!}\right)\left(\sum_{j=0}^{l} \frac{a_{l-j}}{(2j)!}\right)^{-1} \\ &> \left(a_l + \frac{a_{l-1}}{6}\right)\left(a_l + \frac{a_{l-1}}{2} \sum_{j=0}^{l} \frac{1}{8^j}\right)^{-1} \\ &> \left(1 + \frac{1}{6}\frac{a_{l-1}}{a_l}\right)\left(1 + \frac{4}{7}\frac{a_{l-1}}{a_l}\right)^{-1}\end{aligned}$$

and since the function

$$x \to (1 + \tfrac{1}{6}x)(1 + \tfrac{4}{7}x)^{-1}$$

is decreasing, we have

$$a_{K+1}/a_K > (1 + \tfrac{1}{6} \cdot \tfrac{3}{2})(1 + \tfrac{4}{7} \cdot \tfrac{3}{2})^{-1} > \tfrac{2}{3}.$$

Now, if $K = 2l + 1$ is odd, then

$$\frac{a_{K+1}}{a_K} = \frac{a_{2l+2}}{a_{2l+1}} = \left(\sum_{j=0}^{l+1} \frac{a_{l+1-j}}{(2j)!}\right)\left(\sum_{j=0}^{l} \frac{a_{l-j}}{(2j+1)!}\right)^{-1}$$

$$> \left(a_{l+1} + \frac{1}{2}a_l\right)\left(a_l \sum_{j=0}^{l} \frac{1}{4^j}\right)^{-1} > \frac{3}{4}\left(\frac{a_{l+1}}{a_l} + \frac{1}{2}\right) > \frac{2}{3}.$$

and the lemma is proved.

Lemma 8.2.2 If

$$\psi(x) = \exp\left(\sum_{j=0}^{\infty} x^{2^j} - cx^3\right) = \sum_{k=0}^{\infty} b_k x^k$$

and if $c < (\frac{2}{3})^3$, then $b_k > 0$ $(k = 0, 1, 2, \ldots)$.

Proof Indeed, we have

$$\psi(x) = \left(\sum_{l=0}^{\infty} a_l x^l\right)\left(\sum_{m=0}^{\infty} (-1)^m \frac{c^m x^{3m}}{m!}\right),$$

so that

$$b_k = \sum_{l+3m=k} (-1)^m \frac{c^m}{m!} a_l = a_k - a_{k-3}c + a_{k-6}\frac{c^2}{2!} - a_{k-9}\frac{c^3}{3!} + \cdots$$

$$\geq a_{k-3}\left(\frac{a_k}{a_{k-3}} - c\right) + a_{k-9}\frac{c^2}{2!}\left(\frac{a_{k-6}}{a_{k-9}} - \frac{c}{3}\right) + \cdots$$

and from Lemma 8.2.1 we deduce that

$$b_k > a_{k-3}\left(\left(\frac{2}{3}\right)^3 - c\right) + a_{k-9}\frac{c^2}{2!}\left(\left(\frac{2}{3}\right)^3 - \frac{c}{3}\right) + \cdots > 0$$

and the lemma is proved.

Lemma 8.2.3 The function f defined by

$$\log f(t) = \sum_{j=1}^{\infty} \exp(-2^j)(\exp(2^j it) - 1) - ce^{-3}(e^{3it} - 1)$$

is a characteristic function if $c < (\frac{2}{3})^3$.

Proof Indeed,

$$f(t) = (\psi(1))^{-1}\psi(e^{it}/e) = (\psi(1))^{-1}\sum_{k=0}^{\infty} b_k e^{-k}e^{ikt} = \hat{q}(t)$$

where q is the probability concentrated on N and defined by

$$q(\{k\}) = (\psi(1))^{-1}b_k e^{-k} \qquad (k = 0, 1, 2, \ldots). \tag{8.2.10}$$

Proof of Theorem 8.2.3 If $c < (\frac{2}{3})^3$, we deduce from Corollary 4.3.7 that the probability q defined by (8.2.10) divides p and if $c > 0$, from Corollary 4.3.8, q is not infinitely divisible, so that p does not belong to I_0^1.

We note that the probability p defined in Theorem 8.2.3 admits a $(0, 0, \mu)$-De Finetti representation with μ concentrated on a Linnik set $\Lambda \subset Z$ and satisfying

$$\mu(\{\lambda\}) = O(e^{-\lambda})$$

when $\lambda \in \Lambda$, $\lambda \to +\infty$. This proves that condition (8.2.1) is necessary for the validity of Theorem 8.2.2.

We extend now Theorem 8.2.2 to multivariate probabilities.

Theorem 8.2.4 Let $A = \{m_1 e_1, \ldots, m_n e_n\}$ $(m_j \in R, j = 1, \ldots, n)$ and $\hat{p}$ be a characteristic function admitting an $(\alpha, 0, \mu)$-De Finetti representation with $\alpha \in R^n$ and μ a finite signed measure satisfying the two following conditions:

(a) μ is concentrated on a Linnik set $\Lambda \subset (Z)A$;
(b) if $\xi = \sum_{j=1}^n \xi_j m_j e_j \in \Lambda$,

$$|\mu|(\{\xi\}) = o\left(\exp\left(-2\sum_{j=1}^{n} |\xi_j| \log|\xi_j|\right)\right)$$

when $\|\xi\| \to +\infty$.

If q divides p, then $\hat{q}$ admits a $(\beta, 0, \nu)$-De Finetti representation with $\beta \in R^n$ and $\nu \in \mathcal{M}(\Lambda)$.

Proof We can suppose without loss of generality that $\Lambda \subset Z^n$. Then

$$\log \hat{p}(t) = i(\alpha, t) + \sum_{\lambda \in \Lambda} \mu(\{\lambda\})(e^{i(\lambda, t)} - 1) \tag{8.2.11}$$

for any $t \in C^n$ and exactly as in Theorem 8.2.2 we obtain the representation

$$\log \hat{q}(t) = i(\beta, t) + \sum_{\lambda \in Z^n} \nu(\{\lambda\})(e^{i(\lambda, t)} - 1) \tag{8.2.12}$$

for any $t \in C^n$ with

$$\sum_{\lambda \in Z^n} |\nu|(\{\lambda\})e^{-(\lambda, y)} < +\infty \tag{8.2.13}$$

for any $y \in C^n$. Defining p_y, q_y, μ_y, and v_y by

$$\hat{p}_y(t) = \hat{p}(t + iy)/\hat{p}(iy), \qquad \hat{q}_y(t) = \hat{q}(t + iy)/\hat{q}(iy),$$
$$\hat{\mu}_y(t) = \hat{\mu}(t + iy), \qquad \hat{v}_y(t) = \hat{v}(t + iy),$$

we obtain from Corollaries 5.2.2 and 2.5.8 that $\mathrm{pr}_{e_j} q_y$ divides $\mathrm{pr}_{e_j} p_y$ for any $y \in R^n$ and $j = 1, 2, \ldots, n$. But we have the representations

$$\log \widehat{\mathrm{pr}_{e_j} p_y}(t) = i \, \mathrm{pr}_{e_j} \alpha t_j + \sum_{\xi \in \Lambda_j} \mathrm{pr}_{e_j} \mu_y(\{\xi e_j\})(e^{i\xi t_j} - 1)$$

and

$$\log \widehat{\mathrm{pr}_{e_j} q_y}(t) = i \, \mathrm{pr}_{e_j} \beta t_j + \sum_{k=-\infty}^{+\infty} \mathrm{pr}_{e_j} v_y(\{k e_j\})(e^{ikt_j} - 1).$$

From the assumptions, we deduce easily that $\mathrm{pr}_{e_j} p_y$ satisfies the conditions of Theorem 8.2.2 and since $\mathrm{pr}_{e_j} q_y$ divides $\mathrm{pr}_{e_j} p_y$, we have

$$0 \leq \mathrm{pr}_{e_j} v_y(\{\xi e_j\}) \leq \mathrm{pr}_{e_j} \mu_y(\{\xi e_j\})$$

for any $\xi \in Z$. In particular when $\xi \notin \Lambda_j$

$$\mathrm{pr}_{e_j} v_y(\{\xi e_j\}) = \sum_{\{\lambda:\, (\lambda, e_j) = \xi\}} v(\{\lambda\}) e^{-(\lambda, y)} = 0 \tag{8.2.14}$$

for any $y \in R^n$. From (8.2.13), we deduce that the series in (8.2.14) is an entire function of y, so that (8.2.14) holds for any $y \in C^n$. This implies that

$$v(\{\lambda\}) = 0$$

if $\mathrm{pr}_{e_j} \lambda \notin \Lambda_j$ and, since j is arbitrary, v is concentrated on Λ.

Theorem 8.2.5 If, in the preceding theorem, $\xi \in \Lambda$ has an issue with respect to μ, then

$$0 \leq v(\{\xi\}) \leq \mu(\{\xi\}).$$

Proof We can suppose again that $m_j = 1$ $(j = 1, \ldots, n)$. If $\xi = \sum_{j=1}^n \xi_j e_j \in \Lambda$ has an issue with respect to μ, there exist some k and $\omega_1, \ldots, \widehat{\omega_k}, \ldots, \omega_n$ $(\omega_j = \pm 1, j \neq k)$ such that $\xi_k \neq 0$ and

$$|\mu| \left(\left\{ x = \sum_{j=1}^n x_j e_j : x_k = \xi_k, \, \omega_l(x_l - \xi_l) > 0 \text{ for some } l \neq k \right\} \right) = 0.$$

We can write (8.2.11) as

$$\log \hat{p}(t) = i(\alpha, t) + \sum_{\lambda_{k,l} \in \Lambda_k} L_l(t) e^{i\lambda_{k,l} t_k}$$

with

$$L_l(t) = \sum_{\substack{\{x = (x_1, \ldots, x_n): \\ x_k = \lambda_{k,l}\}}} \mu(\{x\}) \exp\left(i \sum_{j \neq k} x_j t_j \right)$$

and (8.2.12) as

$$\log \hat{q}(t) = i(\beta, t) + \sum_{\lambda_{k,l} \in \Lambda_k} M_l(t) e^{i\lambda_{k,l} t_k}$$

with

$$M_l(t) = \sum_{\substack{\{x=(x_1,\dots,x_n):\\ x_k=\lambda_{k,l}\}}} \nu(\{x\}) \exp\left(i \sum_{j \neq k} x_j t_j\right).$$

From Theorem 5.2.2, we have

$$|\log \hat{q}(t)| \leq \exp(o(e^{\|t\|/2}))$$

when $\|t\| \to +\infty$ and since

$$M_l(t) = \int_0^{2\pi} (\log \hat{q}(t) - i(\beta, t)) e^{-i\lambda_{k,l} t_k} \, dt_k,$$

we have

$$|M_l(t)| \leq \exp(o(e^{\|t\|/2}))$$

when $\|t\| \to +\infty$. Since (with the notations of the preceding theorem) $\mathrm{pr}_{e_k} q_y$ divides $\mathrm{pr}_{e_k} p_y$ and since $\mathrm{pr}_{e_k} p_y$ satisfies the assumptions of Theorem 8.2.2, we have

$$0 \leq M_l(iy) \leq L_l(iy) \tag{8.2.15}$$

for any $y \in R^n$.

Now let l_0 be such that $\xi_k = \lambda_{k,l_0}$. We have

$$L_{l_0}(iy) = O(e^{-(\xi, y)})$$

and from (8.2.15)

$$M_{l_0}(iy) = O(e^{-(\xi, y)})$$

for $y \perp e_k$, $\omega_j y_j > 0$, $\|y\| \to +\infty$.

Letting $y_k = 0$, $\omega_j y_j > 0$ $(j \neq k, j \neq l)$ and

$$M_{l_0}(iy) = f(y_l),$$

we have f as an entire function of y_l satisfying

$$|f(y_l)| \leq \exp(o(e^{|y_l|/2})) \tag{8.2.16}$$

when $|y_l| \to +\infty$ and

$$f(y_l) = O(e^{-\xi_l y_l}) \tag{8.2.17}$$

when $\omega_l y_l \to +\infty$. Since

$$f(y_l) = \sum_{\lambda_{l,\alpha} \in \Lambda_l} M_{l,\alpha}(y) \exp(-\lambda_{l,\alpha} y_l)$$

with

$$M_{l,\alpha}(y) = \sum_{\substack{\{x=(x_1,\dots,x_n):\\ x_k=\xi_k,\, x_l=\lambda_{l,\alpha}\}}} \nu(\{x\}) \exp\left(- \sum_{j\neq k,\, j\neq l} x_j y_j \right),$$

from (8.2.16) and (8.2.17)

$$M_{l,\alpha}(y) = 0$$

for $\omega_j y_j > 0$ if $\omega_l(\lambda_{l,\alpha} - \xi_l) > 0$. Since $M_{l,\alpha}$ is an entire function, we have

$$M_{l,\alpha} = 0$$

for any (l, α) satisfying $\omega_l(\lambda_{l,\alpha} - \xi_l) > 0$ and this implies

$$\nu(\{x\}) = 0$$

for any $x = (x_1, \dots, x_n)$ such that $x_k = \xi_k$, $\omega_l(x_l - \xi_l) > 0$. Since this is true for any l, we have

$$|\nu| \left(\left\{ x = \sum_{j=1}^{n} x_j e_j : x_k = \xi_k ,\, \omega_l(x_l - \xi_l) > 0 \text{ for some } l \neq k \right\} \right) = 0.$$

Now

$$L_{l_0}(iy) = (\mu(\{\xi\}) + o(1))e^{-(\xi, y)} \qquad \text{and} \qquad M_{l_0}(iy) = (\nu(\{\xi\}) + o(1))e^{-(\xi, y)}$$

when $\|y\| \to +\infty$, $\omega_j y_j > 0$ $(j = 1, \dots, n; j \neq k)$. From (8.2.15), we obtain

$$0 \leq \nu(\{\xi\}) \leq \mu(\{\xi\})$$

and the theorem is proved.

From Theorems 8.2.4 and 8.2.5, we deduce immediately

Theorem 8.2.6 Let $A = \{m_1 e_1, \dots, m_n e_n\}$ $(m_j \in R, j = 1, \dots, n)$ and $\hat{p}$ be an n-variate characteristic function admitting an $(\alpha, 0, \mu)$-De Finetti representation. We suppose that μ satisfies the following conditions:

(a) μ is concentrated on a Linnik set $\Lambda \subset (Z)A$;
(b) each point of Λ has an issue with respect to μ;
(c) if $\xi = \sum_{j=1}^{n} \xi_j m_j e_j \in \Lambda$,

$$|\mu|(\{\xi\}) = o\left(\exp\left(-2 \sum_{j=1}^{n} |\xi_j| \log |\xi_j| \right) \right)$$

when $\|\xi\| \to +\infty$.

Then p belongs to I_0^n.

With the method used for Corollary 6.4.1, we deduce from this result

Corollary 8.2.2 Let $\lambda_k = (\lambda_{k,1}, \ldots, \lambda_{k,n}) \in R^n$ $(k = 1, 2, \ldots)$ be such that $\lambda_{k',j}/\lambda_{k,j}$ is either a negative rational or an integer greater than one $(k' > k;$ $j = 1, \ldots, n)$ and $l_k = (l^{(k)}_{1,0,\ldots,0}, l^{(k)}_{0,1,0,\ldots,0}, \ldots, l^{(k)}_{1,\ldots,1}) \in (R^+)^{2^n-1}$ be such that

$$\sum_{k=1}^{\infty}\left(\sum_{\varepsilon} l_{\varepsilon}^{(k)}\right) < +\infty$$

$[\varepsilon = (\varepsilon_1, \ldots, \varepsilon_n)$ with $\varepsilon_j = 0$ or 1 and $\sum_\varepsilon$ indicates the summation on the $2^n - 1$ values of $\varepsilon \neq 0]$ and

$$l^{(k)}_{\varepsilon_1,\ldots,\varepsilon_n} = o\left(\exp\left(-2\sum_{j=1}^{n}\varepsilon_j\,|\lambda_{k,j}|\,\log|\lambda_{k,j}|\right)\right)$$

when $\|\lambda_k\| \to +\infty$. Then the sequence of probabilities $\{q_k\}$ defined by

$$q_k = \mathfrak{p}(\lambda_1, l_1) * \cdots * \mathfrak{p}(\lambda_k, l_k)$$

converges completely to a probability p belonging to I_0^n.

8.3 EXTENSIONS TO INDEPENDENT SETS

We extend now the notions of Linnik sets and issue to countable independent sets. Let $A = \{a_j\}$ be a countable independent set of R^n. We say that $\Lambda \subset (Q)A$ is a *Linnik set constructed* on A if

$$\Lambda = \Lambda_1 + \cdots + \Lambda_j + \cdots$$

where $\Lambda_j = \{\lambda_{j,k}\, a_j\}$, $\{\lambda_{j,k}\}$ being a Linnik set of rational numbers. In other words, Λ is the set of all the $l \in R^n$ admitting a representation

$$l = \sum_{j=1}^{\infty} l_j$$

where $l_j \in \Lambda_j$ $(j = 1, 2, \ldots)$, all the l_j being null except a finite number. It must be noted that if A is a basis of R^n, this notion gives only the notion of rational Linnik set given in Chapter 6.

Now let $\xi = \sum_{j=1}^{\infty} \xi_j a_j$ be a point of Λ and $\mu \in \mathscr{M}(\Lambda)$. We say that $\xi \neq 0$ has an *issue with respect to* μ if there exist some k and $\omega_1, \omega_2, \ldots, \widehat{\omega_k}, \ldots$ $(\omega_j = \pm 1, j = 1, 2, \ldots)$ such that $\xi_k \neq 0$ and

$$|\mu|\left(\left\{x = \sum_{j=1}^{\infty} x_j a_j \in \Lambda : x_k = \xi_k,\, \omega_j\,(x_j - \xi_j) > 0 \text{ for some } j \neq k\right\}\right) = 0.$$

We have then

Theorem 8.3.1 Let $\hat{p}$ be a characteristic function admitting an $(\alpha, 0, \mu)$-De Finetti representation. If μ satisfies the following conditions:

(a) μ is concentrated on a Linnik set Λ constructed on a countable independent set $A = \{a_j\}$;
(b) each point of Λ has an issue with respect to μ;
(c) for some $K > 0$

$$|\mu|\left(\left\{x = \sum_{j=1}^{\infty} x_j a_j \in \Lambda : \sum_{j=1}^{\infty} x_j^2 > u^2\right\}\right) = O(\exp(-Ku^2))$$

when $u \to +\infty$,

then p belongs to I_0^n.

Proof We can suppose without loss of generality that $\alpha = 0$ and we consider first the case when A has m elements. If ϕ is the bijection of $(Q)A$ on Q^m defined by

$$\phi\left(\sum_{j=1}^{m} \lambda_j a_j\right) = (\lambda_1, \ldots, \lambda_m)$$

and if Φ is the isomorphism induced on $\mathcal{M}((Q)A)$ by ϕ, then $\Phi(p)$ admits a $(0, 0, \Phi(\mu))$-De Finetti representation. Since $\Phi(\mu)$ satisfies the assumptions of Corollary 8.1.1, $\Phi(p)$ belongs to I_0^m and p belongs to I_0^n.

Now, if A is enumerable, we consider $A_m = \{a_1, \ldots, a_m\}$. Then the projection $P_{A_m}(p)$ admits a $(0, 0, P_{A_m}(\mu))$-De Finetti representation. But $P_{A_m}(\mu)$ is concentrated on the Linnik set $\Lambda_1 + \cdots + \Lambda_m$ and satisfies conditions (b) and (c) of this theorem. It follows from above that $P_{A_m}(p)$ belongs to I_0^n and, from Theorem 7.5.3, p belongs to I_0^n.

With the same method, we deduce from Theorem 8.2.6

Theorem 8.3.2 Let $A = \{a_j\}$ be a countable independent set, $\Lambda \subset (Z)A$ be a Linnik set constructed on A, and μ be a finite signed measure satisfying the following conditions:

(a) μ is concentrated on Λ;
(b) each point of Λ has an issue with respect to μ;
(c) if $\xi = \sum_{j=1}^{\infty} \xi_j a_j \in \Lambda$,

$$|\mu|(\{\xi\}) = o\left(\exp\left(-2\sum_{j=1}^{\infty} |\xi_j| \log|\xi_j|\right)\right)$$

when $\|\xi\| \to +\infty$.

If the n-variate characteristic function $\hat{p}$ admits an $(\alpha, 0, \mu)$-De Finetti representation, then p belongs to I_0^n.

Corollary 8.3.1 Let A be a countable independent set and μ be a finite signed measure concentrated on $A \cup (2)A$. If the characteristic function $\hat{p}$ admits an $(\alpha, 0, \mu)$-De Finetti representation, then p belongs to I_0^n.

Proof $A \cup (2)A$ is contained in the Linnik set

$$\Lambda = \{x = \sum_{j=1}^{n} x_j a_j : x_j = 0, 1, \text{ or } 2\}$$

and if μ is concentrated on $A \cup (2)A$, then each point of Λ has an issue with respect to μ. The result follows then from Theorem 8.3.2, condition (c) of this theorem being trivially satisfied.

We do not know if this result (which extends a particular case of Theorem 7.4.2) is true for noncountable sets.

8.4 FINITE PRODUCTS OF POISSON PROBABILITIES

We consider now a characteristic function $\hat{p}$ admitting an $(\alpha, 0, \mu)$-De Finetti representation with a purely atomic finite signed measure μ having a finite number of atoms. In other words

$$\log \hat{p}(t) = i(\alpha, t) + \sum_{j=1}^{N} \mu_j(e^{i(t, a_j)} - 1) \tag{8.4.1}$$

where $\alpha \in R^n$, $a_j \in R^n$, and $\mu_j \in R$ $(j = 1, \ldots, N)$. If μ_j is positive $(j = 1, \ldots, N)$, p is a finite product of Poisson probabilities. We can find an independent set $B = \{b_1, \ldots, b_m\}$ such that $a_j \in (Z)B$ $(j = 1, \ldots, N)$. Then (8.4.1) can be written in the form

$$\log \hat{p}(t) = i(\alpha, t) + P(e^{i(t, b_1)}, \ldots, e^{i(t, b_m)}) - P(1, \ldots, 1) \tag{8.4.2}$$

where $\alpha \in R^n$ and

$$P(x) = \sum_{r \le k \le s} l_k x^k \tag{8.4.3}$$

for any $x \in C^m$ $(r \in Z^m,\ s \in Z^m,\ r \le s,\ l_k \in R)$.

We have then for the factors of such a probability

Theorem 8.4.1 Let $\{b_1, \ldots, b_m\}$ be an independent set and $\hat{p}$ be the characteristic function defined by (8.4.2) and (8.4.3) with $\alpha \in R^n$, $r \in Z^m$, $s \in Z^m$, $r \le s$, and $l_k \in R$. If q divides p, then

$$\log \hat{q}(t) = i(\beta, t) + Q(e^{i(t, b_1)}, \ldots, e^{i(t, b_m)}) - Q(1, \ldots, 1) \tag{8.4.4}$$

where $\beta \in R^n$ and

$$Q(x) = \sum_{r \leq k \leq s} m_k x^k \tag{8.4.5}$$

for any $x \in C^m$, m_k being real constants.

Proof $\hat{p}$ admits an $(\alpha, 0, \mu)$-De Finetti representation with a finite signed measure μ concentrated on $(Z)B$ and, from Theorem 8.1.1, we deduce that $\hat{q}$ admits a $(\beta, 0, \nu)$-De Finetti representation with a finite signed measure ν concentrated on $(Z)B$. In other words

$$\log \hat{q}(t) = i(\beta, t) + Q(e^{i(t, b_1)}, \ldots, e^{i(t, b_m)}) - Q(1, \ldots, 1)$$

where

$$Q(x) = \sum_{k \in Z^m} m_k x^k,$$

m_k being real constants. We can suppose without loss of generality that $\alpha = \beta = 0$ and we use the bijection ϕ between B and the canonical basis of R^m. If Φ is the isomorphism induced by ϕ on $\mathcal{M}((Z)B)$, $\Phi(p)$ admits a $(0, 0, \Phi(\mu))$-De Finetti representation and $\Phi(\mu)$ is concentrated on the cube $C = \{u \in R^m : r \leq u \leq s\}$. Since $\Phi(q)$ divides $\Phi(p)$ and $\Phi(q)$ admits a $(0, 0, \Phi(\nu))$-De Finetti representation, using the method of Theorem 7.1.1, we obtain that $\Phi(\nu)$ is concentrated on the convex support of $\Phi(\mu)$ which is contained in C. This implies that Q is given by (8.4.5) for any $x \in C^m$ and the theorem is proved.

Corollary 8.4.1 Let $\{b_1, \ldots, b_m\}$ be an independent set and $\hat{p}$ be a characteristic function admitting the representation

$$\hat{p}(t) = e^{i(\alpha, t)} \hat{p}_1(t) \hat{p}_2(t)$$

where $\alpha \in R^n$ and

$$\log \hat{p}_1(t) = P_1(\exp(i(b_1, t)), \ldots, \exp(i(b_{m_1}, t))) - P_1(1, \ldots, 1),$$

$$\log \hat{p}_2(t) = P_2(\exp(i(b_{m_1+1}, t)), \ldots, \exp(i(b_m, t))) - P_2(1, \ldots, 1)$$

with

$$P_j(x_j) = \sum_{r_j \leq k \leq s_j} l_{j, k} x_j^k$$

for $x_j \in C^{m_j}$ ($l_{j, k} \in R$, $r_j \in Z^{m_j}$, $s_j \in Z^{m_j}$, $r_j \leq s_j$, $m_1 + m_2 = m$). Then p_j is a probability ($j = 1, 2$) and if q divides p, then

$$\hat{q}(t) = e^{i(\beta, t)} \hat{q}_1(t) \hat{q}_2(t)$$

where $\beta \in R^n$ and

$$\log \hat{q}_1(t) = Q_1(\exp(i(b_1, t)), \ldots, \exp(i(b_{m_1}, t))) - Q_1(1, \ldots, 1),$$

$$\log \hat{q}_2(t) = Q_2(\exp(i(b_{m_1+1}, t)), \ldots, \exp(i(b_m, t))) - Q_2(1, \ldots, 1)$$

with

$$Q_j(x_j) = \sum_{r_j \le k \le s_j} m_{j,k} x_j^k.$$

Moreover, q_j is a probability dividing p_j $(j = 1, 2)$.

Proof This follows immediately from the preceding theorem and Lemma 7.4.2.

Theorem 8.4.1 proves the importance of the functions f which admit a representation

$$\log f(t) = Q(e^{i(t,b_1)}, \ldots, e^{i(t,b_m)}) - Q(1, \ldots, 1) \tag{8.4.6}$$

where Q is defined by (8.4.5) and which are characteristic functions. We can characterize these functions by

Theorem 8.4.2 Let Q and f be defined, respectively, by (8.4.5) and (8.4.6) with $b_j \in R^n$ and $m_k \in R$. If all the coefficients q_k of the expansion of e^Q in Laurent series

$$e^{Q(x)} = \sum_{k \in Z^m} q_k x^k \tag{8.4.7}$$

are nonnegative, f is the characteristic function of an n-variate probability. Conversely, if f is the characteristic function of an n-variate probability and if $\{b_1, \ldots, b_m\}$ is an independent set, all the coefficients q_k in (8.4.7) are nonnegative.

Proof If $C = \exp(-Q(1, \ldots, 1))$, then

$$f(t) = C \exp(Q(e^{i(t,b_1)}, \ldots, e^{i(t,b_m)})$$
$$= C \sum_{k=(k_1,\ldots,k_m) \in Z^m} q_k \exp\left(i\left(t, \sum_{j=1}^m k_j b_j\right)\right)$$

and therefore f is the Fourier–Stieltjes transform of the signed measure q defined by

$$q(B) = C \sum_{k_1 b_1 + \cdots + k_m b_m \in B} q_k$$

for any $B \in \mathscr{B}_n$. If all the q_k are nonnegative, then $q(B) \ge 0$ for any $B \in \mathscr{B}_n$ and q is a probability. Conversely, if $\{b_1, \ldots, b_m\}$ is an independent set, then

$$q(\{k_1 b_1 + \cdots + k_m b_m\}) = q_{k_1,\ldots,k_m}$$

and if q is a probability, then $q_k \ge 0$ for any $k \in Z^m$.

The functions Q defined by (8.4.5) which have nonnegative coefficients q_k in the expansion (8.4.7) are studied in Appendix B. We give only the final result and for a univariate function f

$$f(x) = \sum_{j=-p}^{q} a_j x^j$$

($a_j \in R$), we introduce for any j ($-p \le j \le q, j \ne 0$) the sets

$$\mathfrak{A}_j = \{k : a_k \ne 0,\, k/j > 1\},$$

$$\mathfrak{B}_j = \{k : a_k \ne 0,\, 0 < k/j < 1\},$$

$$\mathfrak{C}_j = \{\,|k| : a_k \ne 0,\, k/j < 0\},$$

and the following conditions.

Condition A For any $a_j < 0$, $j \in (Z)\mathfrak{A}_j$.

Condition B For any $a_j < 0$, $j \in (N)\mathfrak{B}_j$.

Condition C For any $a_j < 0$, $j \in (N)(\mathfrak{B}_j \cup \{\operatorname{sgn} j\rho_j\})$ where ρ_j is the g.c.d. of $\mathfrak{C}_j$ (if $\mathfrak{C}_j = \varnothing$, then $\rho_j = 0$).

We have then

Theorem 8.4.3 Let Q be the function defined by (8.4.5) where the m_k are real constants. In expansion (8.4.7), all the coefficients q_k are nonnegative if and only if the two following conditions are satisfied.

(a) The function f defined by

$$f(y) = Q(y^{p_1}, \ldots, y^{p_m}) \tag{8.4.8}$$

satisfies Conditions A and C for any integers $p_1, \ldots, p_m$.

(b) The absolute values of negative coefficients m_k are small enough.

If Q is a polynomial, condition (a) can be replaced by

(a′) the function f defined by (8.4.8) satisfies Conditions A and B for any nonnegative integers $p_1, \ldots, p_m$.

From this result, we deduce the following theorem which solves the problem of the belonging to I_0^n of finite products of Poisson laws.

Theorem 8.4.4 Let $\{b_1, \ldots, b_m\}$ be an independent set of R^n and let $\hat{p}$ be the characteristic function defined by (8.4.2) and (8.4.3) with $\alpha \in R^n$, $r \in Z^m$, $s \in Z^m$, $r \le s$, and $l_k \ge 0$. Then p does not belong to I_0^n if and only if there exists some $k_0 \in Z^m$ ($k_0 \ne 0$, $r \le k_0 \le s$) such that the function f defined by

$$f(y) = Q(y^{p_1}, \ldots, y^{p_m}), \qquad Q(x) = P(x) - (l_{k_0} + \varepsilon)x^{k_0} \tag{8.4.9}$$

satisfies Conditions A and C for some $\varepsilon > 0$ and any integers $p_1, \ldots, p_m$.

When the function P defined by (8.4.3) is a polynomial, this last condition can be replaced by the following one. There exists some $k_0 \in Z^m$ ($k_0 \neq 0$, $r \leq k_0 \leq s$) such that the function f defined by (8.4.9) satisfies Conditions A and B for some $\varepsilon > 0$ and any nonnegative integers $p_1, \ldots, p_m$.

Proof Let Q and f be defined by (8.4.9). If for some $\varepsilon > 0$ the function f satisfies Conditions A and C for any integers $p_1, \ldots, p_m$ and if ε is small enough, it follows from Theorem 8.4.3 that all the coefficients of the expansion of e^Q in Laurent series are nonnegative and from Theorem 8.4.2 the function g defined by

$$\log g(t) = Q(e^{i(t, b_1)}, \ldots, e^{i(t, b_m)}) - Q(1, \ldots, 1)$$

is the characteristic function of a probability q. From Corollary 4.3.7, we deduce that q divides p and, since Q has a negative coefficient, q is not infinitely divisible, so that p does not belong to I_0^n.

We prove now the converse in the case $m = 1$. Then p is defined by

$$\log \hat{p}(t) = i(\alpha, t) + P(e^{ibt}) - P(1)$$

where

$$P(x) = \sum_{j=p}^{q} a_j x^j$$

($a_j \geq 0$). If

$$p = p' * p'', \tag{8.4.10}$$

we deduce from Theorem 8.4.1 that

$$\log \hat{p}'(t) = i(\alpha', t) + P'(e^{ibt}) - P'(1)$$

where

$$P'(x) = \sum_{j=p}^{q} a'_j x^j$$

and

$$\log \hat{p}''(t) = i(\alpha'', t) + P''(e^{ibt}) - P''(1)$$

where

$$P''(x) = \sum_{j=p}^{q} a''_j x^j.$$

From (8.4.10), we deduce easily the relation

$$P = P' + P''$$

or

$$a_j = a'_j + a''_j \qquad (j = p, p + 1, \ldots, q). \tag{8.4.11}$$

We define the sets $\mathfrak{A}_j$, $\mathfrak{B}_j$, and $\mathfrak{C}_j$ by

$$\mathfrak{A}_j = \{k : a_k \neq 0, k/j > 1\},$$
$$\mathfrak{B}_j = \{k : a_k \neq 0, 0 < k/j < 1\},$$
$$\mathfrak{C}_j = \{|k| : a_k \neq 0, k/j < 0\},$$

and the sets $\mathfrak{A}'_j$, $\mathfrak{B}'_j$, $\mathfrak{C}'_j$, $\mathfrak{A}''_j$, $\mathfrak{B}''_j$, and $\mathfrak{C}''_j$ with obvious notations. If p does not belong to I_0^n, there exists a decomposition (8.4.10) with at least one of the probabilities p' or p'' which is not infinitely divisible. We can suppose that it is p'. Then one of the a'_j $(j \neq 0)$ is negative. We can suppose without loss of generality that $j > 0$. Since p' is a probability, it follows from Theorems 8.4.2 and 8.4.3 that we have the two following conditions:

(a) $j \in (Z)\mathfrak{A}'_j$;
(b) $j \in (N)(\mathfrak{B}'_j \cup \{\operatorname{sgn} j\rho'_j\})$ where ρ'_j is the g.c.d. of $\mathfrak{C}'_j$.

We prove first that

$$(Z)\mathfrak{A}'_j \subset (Z)\mathfrak{A}_j. \tag{8.4.12}$$

If $\mathfrak{A}'_j \subset \mathfrak{A}_j$, then (8.4.12) is clear. Otherwise, let

$$\mathfrak{A}'_j \backslash \mathfrak{A}_j = \{k_1, \ldots, k_\beta\}$$

$(|k_1| > \cdots > |k_\beta|)$. We have then

$$a'_{k_1} \neq 0, \qquad a_{k_1} = 0.$$

From (8.4.9), we deduce that one of the a'_{k_1} or a''_{k_1} is negative. If $a'_{k_1} < 0$, then $k_1 \in (Z)\mathfrak{A}'_{k_1}$ and since $\mathfrak{A}'_{k_1} \subset \mathfrak{A}_{k_1}$,

$$k_1 \in (Z)\mathfrak{A}_{k_1}.$$

In the same way, if $a''_{k_1} < 0$, we obtain that $k_1 \in (Z)\mathfrak{A}_{k_1}$. We can iterate this procedure to $k_2, \ldots, k_\beta$ and obtain

$$k_l \in (Z)\mathfrak{A}_{k_1}$$

$(l = 1, \ldots, \beta)$ and this implies (8.4.12).

We prove now that

$$(N)(\mathfrak{B}'_j \cup \{\operatorname{sgn} j\rho'_j\}) \subset (N)(\mathfrak{B}_j \cup \{\operatorname{sgn} j\rho_j\}), \tag{8.4.13}$$

where ρ_j is the g.c.d. of $\mathfrak{C}_j$. With the method used above, we obtain that ρ'_j is a multiple of ρ_j. If $\mathfrak{B}'_j \subset \mathfrak{B}_j$, (8.4.13) is clear. Otherwise, let

$$\mathfrak{B}'_j \backslash \mathfrak{B}_j = \{k'_1, \ldots, k'_\gamma\}$$

$(|k'_1| < \cdots < |k'_\gamma|)$. We have then

$$a'_{k_1'} \neq 0, \qquad a_{k_1'} = 0.$$

From (8.4.11), we deduce that one of the $a'_{k_1'}$ or $a''_{k_1'}$ is negative. If $a'_{k_1'} < 0$, then $k'_1 \in (N)(\mathfrak{B}'_{k_1'} \cup [\operatorname{sgn} j\rho'_j\})$ and, since ρ'_j is a multiple of ρ_j and $\mathfrak{B}'_{k_1'} \subset \mathfrak{B}_{k_1'}$, then

$$k'_1 \in (N)(\mathfrak{B}_{k_1'} \cup \{\operatorname{sgn} j\rho_j\}). \tag{8.4.14}$$

If $a''_{k_1'} < 0$, we prove (8.4.14) with an analogous method. We can iterate this procedure to $k'_2, \ldots, k'_\gamma$ and obtain

$$k'_l \in (N)(\mathfrak{B}_{k_1'} \cup \{\operatorname{sgn} j\rho_j\})$$

$(l = 1, \ldots, \gamma)$ and this proves (8.4.13). From (8.4.12) and (8.4.13), we deduce that the function Q defined by

$$Q(x) = P(x) - (a_j + \varepsilon)x^j$$

satisfies Conditions A and C for any $\varepsilon > 0$.

For the general case, we can suppose $\alpha = 0$ and let ϕ be the bijection between $B = \{b_1, \ldots, b_m\}$ and the canonical basis of R^m. If Φ is the isomorphism induced by ϕ on $\mathscr{M}((Z)B)$, then

$$\log \Phi(p)(u) = P(e^{iu_1}, \ldots, e^{iu_m}) - P(1, \ldots, 1)$$

for any $u = (u_1, \ldots, u_m) \in R^m$. If

$$p = p' * p'',$$

we can suppose that p' and p'' both belong to $\mathscr{P}((Z)B)$ and this implies

$$\Phi(p) = \Phi(p') * \Phi(p'').$$

From this, we deduce that

$$\operatorname{pr}_\theta \Phi(p) = \operatorname{pr}_\theta \Phi(p') * \operatorname{pr}_\theta \Phi(p'')$$

for any rational vector θ of R^m and the converse follows easily from this relation and the case $m = 1$.

We give now some applications of this result. Most of these are based on the following.

Lemma 8.4.1 Let f be the function defined by

$$f(y) = \sum_{j=p}^{q} m_j y^j$$

$(y \in C)$. If f satisfies Conditions A and C and if one of the coefficients m_j is negative, then at least three coefficients m_j are positive.

Proof Let $L = \{j : j \neq 0, m_j < 0\}$ and $M = \{j : j \neq 0, m_j > 0\}$. Let j_0 and j'_0 be, respectively, the greatest and the smallest element of L. If $0 < j'_0 \leq j_0$ (resp. $j'_0 \leq j_0 < 0$), applying Condition A to j_0 (resp. to j'_0), we obtain the existence of two elements of M greater than j_0 (resp. smaller than j'_0) and applying Condition B to j'_0 (resp. to j_0), we obtain the existence of one element of M smaller than j'_0 (resp. greater than j_0). If $j'_0 < 0 < j_0$, applying Condition A to j_0 and j'_0, we obtain the existence of two elements of M greater than j_0 and of two elements of M smaller than j'_0. In all the cases, the lemma is proved.

Theorem 8.4.5 If a characteristic function $\hat{p}$ admits an $(\alpha, 0, \mu)$-De Finetti representation with a purely atomic measure μ having two atoms, then p belongs to I_0^n.

Proof Let a_1 and a_2 be the two atoms of μ. If $\{a_1, a_2\}$ is an independent set, then the theorem follows immediately from Theorem 7.4.2. If $\{a_1, a_2\}$ is not an independent set, there exists some $b \in R^n$ such that

$$\log \hat{p}(t) = i(\alpha, t) + P(e^{i(b, t)}) - P(1)$$

where

$$P(x) = l_1 x^p + l_2 x^q$$

$(p, q \in Z)$. From Lemma 8.4.1, we deduce that the function Q defined by

$$Q(x) = P(x) - (m_k + \varepsilon)x^k$$

($m_k = l_1$ for $k = p$, l_2 for $k = q$, 0 otherwise) does not satisfy Conditions A and C for any $\varepsilon > 0$ and, from Theorem 8.4.4, p belongs to I_0^n.

For the case $n = 1$, we obtain immediately from this result

Corollary 8.4.2 The product of two univariate Poisson probabilities belongs to I_0^n.

We prove now

Theorem 8.4.6 Let $\hat{p}$ be a characteristic function admitting an $(\alpha, 0, \mu)$-De Finetti representation with a purely atomic finite signed measure μ having three atoms. If p does not belong to I_0^n, there exist some $b \in R^n$ and some integers $\alpha_1, \alpha_2, \alpha_3$ satisfying one of the following conditions:

(a) $0 < 2\alpha_1 < \alpha_2 < \alpha_3$;
(b) $0 < -\alpha_1 < \alpha_2 < \alpha_3$,

such that $a_j = \alpha_j b$ is an atom of μ $(j = 1, 2, 3)$.

Proof Let a_j be an atom of μ $(j = 1, 2, 3)$. If $\{a_1, a_2, a_3\}$ is an independent set, it follows from Theorem 7.4.2 that p belongs to I_0^n.

We suppose now the existence of an independent set $\{b_1, b_2\}$ such that

$$a_1 = \alpha b_1, \qquad a_2 = \beta b_2, \qquad a_3 = \gamma b_1 + \delta b_2 \tag{8.4.15}$$

where α, β, γ, and δ are integers. Then

$$\log \hat{p}(t) = P(e^{i(b_1, t)}, e^{i(b_2, t)}) - P(1, 1)$$

where

$$P(x_1, x_2) = l_1 x_1^\alpha + l_2 x_2^\beta + l_3 x_1^\gamma x_2^\delta .$$

If $\gamma = 0$ (or $\delta = 0$), it follows from Corollary 8.4.1 and Theorem 8.4.5 that $\hat{p}$ belongs to I_0^n. If $\gamma \neq 0$ and $\delta \neq 0$ and if p does not belong to I_0^n, we obtain from Theorem 8.4.4 the existence of some $k = (k_1, k_2) \in R^2$ $(k \neq 0)$ such that the function f defined by

$$Q(x) = P(x) - (m_k + \varepsilon)x^k, \qquad f(y) = Q(y^{p_1}, y^{p_2})$$

[$m_k = l_1$ if $k = (\alpha, 0)$, l_2 if $k = (0, \beta)$, l_3 if $k = (\gamma, \delta)$, 0 otherwise] satisfies Conditions A and C for some $\varepsilon > 0$ and any integers p_1 and p_2. From Lemma 8.4.1, we deduce immediately that

$$k \notin \{(\alpha, 0), (0, \beta), (\gamma, \delta)\}. \tag{8.4.16}$$

If $p_1 = 1$, $p_2 = 0$, then

$$f(y) = l_1 y^\alpha + l_3 y^\gamma - \varepsilon y^{k_1}$$

and from Lemma 8.4.1, f does not satisfy Conditions A and C if $k_1 \notin \{0, \alpha, \gamma\}$. In the same way, the case $p_1 = 0$, $p_2 = 1$ implies $k_2 \in \{0, \beta, \delta\}$. Therefore, k is one of the five points: $(\gamma, 0)$, $(0, \delta)$, (α, β), (γ, β), (α, δ).

If $p_1 = \delta$, $p_2 = -\gamma$, then

$$f(y) = l_1 y^{\alpha\delta} + l_2 y^{-\beta\gamma} - \varepsilon y^{k_1\delta - k_2\gamma}$$

and, from Lemma 8.4.1,

$$k_1\delta - k_2\gamma \in \{0, \alpha\delta, -\beta\gamma\}. \tag{8.4.17}$$

If $p_1 = \beta$, $p_2 = \alpha$, then

$$f(y) = (l_1 + l_2) y^{\alpha\beta} + l_3 y^{\beta\gamma + \alpha\delta} - \varepsilon y^{k_1\beta + k_2\alpha}$$

and, from Lemma 8.4.1,

$$k_1\beta + k_2\alpha \in \{0, \alpha\beta, \beta\gamma + \alpha\delta\}. \tag{8.4.18}$$

If $p_1 = \delta$, $p_2 = \alpha - \gamma$, then

$$f(y) = (l_1 + l_3) y^{\alpha\delta} + l_2 y^{(\alpha-\gamma)\beta} - \varepsilon y^{k_1\delta + k_2(\alpha-\gamma)}$$

and from Lemma 8.4.1

$$k_1\delta + k_2(\alpha - \gamma) \in \{0, \alpha\delta, (\alpha - \gamma)\beta\}. \tag{8.4.19}$$

If $k = (\gamma, 0)$, from (8.4.18), we have $\beta\gamma \in \{0, \alpha\beta, \beta\gamma + \alpha\delta\}$ but $\beta\gamma = 0$ and $\beta\gamma = \beta\gamma + \alpha\delta$ are impossible while $\beta\gamma = \alpha\beta$ gives $k = (\alpha, 0)$ which contradicts (8.4.16). Therefore $k \neq (\gamma, 0)$ and, by symmetry, $k \neq (0, \delta)$. If $k = (\alpha, \beta)$, from (8.4.17), we have $\alpha\delta - \beta\gamma \in \{0, \alpha\delta, -\beta\gamma\}$ and this implies $\alpha\delta = \beta\gamma$ and, from (8.4.18), we have $2\alpha\beta \in \{0, \alpha\beta, \beta\gamma + \alpha\delta\}$ and this implies $2\alpha\beta = \beta\gamma + \alpha\delta$, so that $\alpha = \gamma$ and $\beta = \delta$ and this contradicts (8.4.16). If $k = (\gamma, \beta)$, from (8.4.17), we obtain $\gamma\delta - \beta\gamma \in \{0, \alpha\delta, -\beta\gamma\}$ but $\gamma\delta - \beta\gamma = -\beta\gamma$ is impossible, while $\gamma\delta - \beta\gamma = 0$ implies $\beta = \delta$ and this contradicts (8.4.16). We have then

$$\gamma\delta - \beta\gamma = \alpha\delta. \tag{8.4.20}$$

From (8.4.18), we have $\beta\gamma + \alpha\beta \in \{0, \alpha\beta, \beta\gamma + \alpha\delta\}$ and this implies $\beta\gamma + \alpha\beta = 0$, so that

$$\alpha = -\gamma \tag{8.4.21}$$

and with (8.4.20), we obtain

$$\beta = 2\delta. \tag{8.4.22}$$

From (8.4.19), we have $\gamma\delta + \beta(\alpha - \gamma) \in \{0, \alpha\delta, (\alpha - \gamma)\beta\}$ which has two solutions: $\gamma\delta + \beta(\alpha - \gamma) = \alpha\delta$ and $\gamma\delta + \beta(\alpha - \gamma) = 0$. The first solution, with (8.4.21), implies $\beta = \delta$ which contradicts (8.4.22) and the second one with (8.4.21) implies $2\beta = \delta$ which also contradicts (8.4.22). Therefore, $k = (\gamma, \beta)$ is impossible and, by symmetry, $k = (\alpha, \delta)$ is also impossible. Therefore, if p does not belong to I_0^n, the atoms a_1, a_2, and a_3 do not satisfy (8.4.15) for some independent set $\{b_1, b_2\}$.

This implies the existence of some $b \in R^n$ and some $\alpha_j \in Z$ such that $a_j = \alpha_j b$ $(j = 1, 2, 3)$. We can enumerate the a_j so that one of the conditions

$$0 < \alpha_1 < \alpha_2 < \alpha_3, \qquad \alpha_1 < 0 < \alpha_2 < \alpha_3$$

is satisfied. Then

$$\log \hat{p}(t) = P(e^{ibt}) - P(1)$$

with

$$P(x) = \sum_{j=1}^{3} l_j x^{\alpha_j}.$$

If p does not belong to I_0^n, there exists some $k \neq 0$ such that the function Q defined by

$$Q(x) = P(x) - (m_k + \varepsilon)x^k$$

($m_k = l_j$ if $k = \alpha_j$, 0 otherwise) satisfies Conditions A and C. From Condition C, we deduce that $k < \alpha_2$ while Condition A implies $k \geq 2\alpha_1$ in the first case and $k \geq -\alpha_1$ in the second one. The theorem is proved.

From Theorem 8.4.4, we deduce easily that the probabilities p_1 and p_2 defined by

$$\log \hat{p}_j(t) = P_j(e^{i(b,\,t)}) - P_j(1)$$

where $b \in R^n$ and

$$P_1(x) = l_1 x^k + l_2 x^{2k+1} + l_3 x^{2k+2},$$

$$P_2(x) = l'_1 x^{-k} + l'_2 x^{k+1} + l'_3 x^{k+2}$$

$(k \in N, l_j > 0, l'_j > 0)$ do not belong to I_0^n. These examples satisfy exactly the condition of Theorem 8.4.6.

We consider now the bivariate characteristic function $\hat{p}$ defined by

$$\log \hat{p}(t) = i(\alpha, t) + \int_A (e^{i(t,\,x)} - 1)\lambda(dx) + l_1 e^{i(t,\,e_1)} + l_2 e^{i(t,\,3e_2)} + l_3 e^{i(t,\,e_1+3e_2)} \tag{8.4.23}$$

where $\alpha \in R^n$, $l_j > 0$, and $A = \{x : 0 < x_1 < 2m,\ 0 < x_2 < 2m,\ m < \mathrm{pr}_{e_1+e_2} x\}$ $(0 < m < \frac{1}{2})$. It follows from Theorem 7.1.7 that p belongs to I_0^2 and from Theorem 7.2.2 that $\mathrm{pr}_\theta\, p$ does not belong to I_0^2 for any $\theta \neq k(e_1 + e_2)$ $(k \in R)$. Now if $\theta = e_1 + e_2$, the probability q defined by

$$\log \hat{q}(t) = l_1 e^{i(t,\,\theta)} + l_2 e^{3i(t,\,\theta)} + l_3 e^{4i(t,\,\theta)}$$

divides $\mathrm{pr}_\theta\, p$ and, from the preceding example, q has an indecomposable factor. Therefore the probability p defined by (8.4.23) belongs to I_0^2, but $\mathrm{pr}_\theta\, p$ does not belong to I_0^2 for any $\theta \in R^2$.

For the Poisson measures having four atoms, we deduce immediately from Corollary 8.4.1 and Theorem 8.4.5

Theorem 8.4.7 Let $\hat{p}$ be an n-variate characteristic function admitting an $(\alpha, 0, \mu)$-De Finetti representation with a purely atomic finite signed measure μ having four atoms: a_j $(1 \leq j \leq 4)$. If there exists some independent set $\{b_1, b_2\}$ and some integers α_j $(1 \leq j \leq 4)$ such that

$$a_1 = \alpha_1 b_1, \qquad a_2 = \alpha_2 b_1, \qquad a_3 = \alpha_3 b_2, \qquad a_4 = \alpha_4 b_2,$$

then p belongs to I_0^n.

Finally, we prove

Theorem 8.4.8 Let $\{b_1, b_2\}$ be an independent set and $\hat{p}$ be the characteristic function defined by

$$\log \hat{p}(t) = P(e^{i(b_1,\,t)}, e^{i(b_2,\,t)}) - P(1, 1)$$

where

$$P(x_1, x_2) = l_1 x_1^m + l_2 x_2^m + l_3 x_1^{2m} + l_4 x_2^{2m} + l_5 x_1^{2m-1} x_2 + l_6 x_2^{2m-1} x_1.$$

If $l_j > 0$ $(1 \leq j \leq 6)$ and m is an integer greater than one, then p does not belong to I_0^n.

Proof If Q is the function

$$Q(x_1, x_2) = P(x_1, x_2) - \varepsilon x_1^m x_2^m,$$

the function f defined by

$$f(y) = Q(y^{p_1}, y^{p_2})$$

satisfies Conditions A and C for any integers p_1 and p_2 and some $\varepsilon > 0$. From Theorem 8.4.4, p does not belong to I_0^n.

From this example, we deduce that Corollary 7.4.1 is not true for a type greater than 2.

NOTES

The theorems of Sections 8.1 and 8.3 generalize some results due to Rousseau (1973, 1974). Theorem 8.2.2 is due to Ostrovskiĭ (1964) while Theorem 8.2.3 is taken from Goldberg and Ostrovskiĭ (1967). Theorems 8.2.4–8.2.8 are new but Corollary 8.2.2 is due to Cuppens (1967). The results on finite products of Poisson laws are due to Lévy (1937a, 1938b) for the case $m = n = 1$ (with our notations) and to Cuppens (1970) in the general case.

Chapter 9

α-Decompositions

9.1 STATEMENT OF THE PROBLEM

Let $\hat{p}$ be an n-variate characteristic function without zeros. We say that the probability q is an *α-factor* of p if there exist some probabilities $r_1, \ldots, r_m$ and some positive constants $\alpha, \beta_1, \ldots, \beta_m$ such that†

$$\hat{p}(t) = \hat{q}^{\alpha}(t) \prod_{j=1}^{m} \hat{r}_j^{\beta_j}(t) \tag{9.1.1}$$

for any $t \in R^n$. Equation (9.1.1) is an *α-decomposition* of $\hat{p}$.

More generally, if $\hat{p}$ is an n-variate characteristic function without zeros, q is an *enumerable α-factor* of p if there exist a sequence of probabilities $\{r_j\}$ and positive constants α and $\{\beta_j\}$ such that

$$\hat{p}(t) = \hat{q}^{\alpha}(t) \prod_{j=1}^{\infty} \hat{r}_j^{\beta_j}(t) \tag{9.1.2}$$

for any $t \in R^n$. Equation (9.1.2) is an *enumerable α-decomposition* of p.

It is clear that any factor of a probability p is an α-factor of p and that any α-factor of p is an enumerable α-factor of p, but a probability can have α-factors which are not factors. For example, if $\hat{p}$ has no zeros, p^{*k} is an

† Here $\hat{q}^{\alpha} = \exp(\alpha \log \hat{q})$, $\hat{r}^{\beta_j} = \exp(\beta_j \log \hat{r})$.

α-factor of p, but does not divide p if $p \neq \delta_m$. A less trivial example is given by the following example. Let p, q, and r be the univariate probabilities defined by

$$p = (1/35)(\delta_0 + 2\delta_1 + 5\delta_2 + 12\delta_3 + 15\delta_4),$$
$$q = (1/7)(\delta_0 + 3\delta_1 + 3\delta_2),$$
$$r = (1/175)(\delta_0 + \delta_1 + 8\delta_2 + 17\delta_3 + 28\delta_4 + 45\delta_5 + 75\delta_6).$$

Since

$$\begin{aligned}\hat{p}(t) &= (1/35)(1 + 2e^{it} + 5e^{2it} + 12e^{3it} + 15e^{4it})\\ &= (1/35)(1 - e^{it} + 5e^{2it})(1 + 3e^{it} + 3e^{2it}),\end{aligned}$$

we prove easily that $\hat{p}$ has no real zeros and that p is indecomposable. From

$$p * p = q * r,$$

we deduce that p which is indecomposable has for α-factors q and r.†

The problem of the description of the α-factors (or the enumerable α-factors) of a probability is a generalization of the one of the description of its factors which appears naturally in some problems of characterization of probabilities [see Lukacs and Laha (1964), Ramachandran (1965, Chapter 8), or Kagan *et al.* (1972)].

This problem is a very difficult one because we do not know if there exists between the supports of the probabilities p, q, and r_j satisfying (9.1.1) or (9.1.2) a relation similar to the one given by Corollary 2.5.2 for ordinary decompositions. Nevertheless, we can prove a more specialized result.

Theorem 9.1.1 If p is an n-variate probability concentrated on an enumerable set A and having a characteristic function without zeros and if q is an enumerable α-factor of p, there exists some $m \in R^n$ such that q is concentrated on $(Z)A + m$.

Proof We suppose first that p and q are both symmetric. Then, from (9.1.2), we have

$$1 - \hat{q}(h) \leq 1 - \hat{p}^{1/\alpha}(h)$$

for any $h \in R^n$ and from Theorem 2.1.4, we have

$$|\hat{q}(t + h) - \hat{q}(t)|^2 \leq 2(1 - \hat{q}(h)).$$

From these two inequalities and from the relations

$$1 - d^l \leq l(1 - d)$$

† It can be proved that q and r are also indecomposable.

valid for $0 \leq d \leq 1$ and $l \geq 1$ and

$$1 - d^l \leq 1 - d$$

valid for $0 \leq d \leq 1$ and $l \leq 1$, we obtain

$$|\hat{q}(t + h) - \hat{q}(t)|^2 \leq C(1 - \hat{p}(h))$$

where C is equal to $2/\alpha$ if $\alpha \leq 1$ and to 2 if $\alpha \geq 1$. Since $\hat{p}$ is an almost periodic function, this implies that $\hat{q}$ is also an almost periodic function and that every ε-almost period of $\hat{p}$ is a $(C\varepsilon)^{1/2}$-almost period of $\hat{q}$. We deduce then from Theorem A.2.4 that q is concentrated on $(Z)A$.

If $\hat{p}$ and $\hat{q}$ are arbitrary, then $|\hat{q}|^2$ is an enumerable α-factor of $|\hat{p}|^2$. Since the probability of $|\hat{p}|^2$ is concentrated on $(Z)A$, then the probability of $|\hat{q}|^2 = \hat{q} \cdot \hat{\bar{q}}$ is concentrated on $(Z)A$ and, from Corollary 2.5.3, we deduce the theorem.

9.2 α-DECOMPOSITIONS OF PROBABILITIES WITH ANALYTIC CHARACTERISTIC FUNCTIONS

Theorem 9.2.1 Let Γ be an open convex set containing the origin and $\hat{p}$ be a characteristic function which is regular in the convex tube $R^n + i\Gamma$, α_j be some positive constants $(j = 1, \ldots, m)$, and q_j be some probabilities $(j = 1, \ldots, m)$. If there exists some $\rho > 0$ such that the relation

$$\hat{p}(t) = \prod_{j=1}^{m} \hat{q}_j^{\alpha_j}(t) \tag{9.2.1}$$

holds for $\|t\| < \rho$, then $\hat{q}_j$ is regular in $R^n + i\Gamma$ and relation (9.2.1) holds in any domain of $R^n + i\Gamma$ containing the origin where $\hat{p}$ does not vanish.

Proof For the case $n = 1$, see Lukacs (1970, Theorem 10.1.1). The general case follows from the case $n = 1$ exactly as Theorem 5.3.1 follows from the Raĭkov's theorem.

From the ridge property, we deduce easily the following corollaries.

Corollary 9.2.1 With the conditions of the preceding theorem, we have

$$|\hat{p}(x + iy)/\hat{p}(iy)|^{1/\alpha_j} \leq |\hat{q}_j(x + iy))/\hat{q}_j(iy)| \leq 1$$

for any $x \in R^n$, $y \in \Gamma$ $(j = 1, \ldots, m)$.

Corollary 9.2.2 If in Theorem 9.2.1, $\hat{p}$ has no zeros in $R^n + i\Gamma$, then

$$0 \leq \operatorname{Re}(\log \hat{q}_j(iy) - \log \hat{q}_j(x + iy)) \leq (1/\alpha_j) \operatorname{Re}(\log \hat{p}(iy) - \log \hat{p}(x + iy))$$

for any $x \in R^n$, $y \in \Gamma$ $(j = 1, \ldots, m)$.

From the corresponding univariate result [cf. Ramachandran (1967, Theorem 7.4)], we can extend Theorem 9.2.1 to enumerable α-decompositions as following.

Theorem 9.2.2 Let Γ be an open convex set containing the origin, $\hat{p}$ be a characteristic function which is regular in the convex tube $R^n + i\Gamma$, $\{q_j\}$ be a sequence of probabilities, and $\{\alpha_j\}$ be a sequence of positive numbers satisfying

$$\alpha_j \geq \varepsilon > 0 \tag{9.2.2}$$

($j = 1, 2, \ldots$). If there exists some $\rho > 0$ such that the relation

$$\hat{p}(t) = \prod_{j=1}^{\infty} \hat{q}_j^{\alpha_j}(t) \tag{9.2.3}$$

holds for $\|t\| < \rho$, then $\hat{q}_j$ is regular in $R^n + i\Gamma$ and relation (9.2.3) holds in any domain of $R^n + i\Gamma$ containing the origin where $\hat{p}$ does not vanish.

From this result and the ridge property, we can obtain the analogs of Corollaries 9.2.1 and 9.2.2 for enumerable α-decompositions.

We do not know if condition (9.2.2) can be omitted in Theorem 9.2.2. Nevertheless, we can weaken the hypothesis of this theorem as following (Theorem 9.2.1 admits evidently a similar extension).

Theorem 9.2.3 Let Γ be an open convex set containing the origin, $\hat{p}$ be a characteristic function regular in the convex tube $R^n + i\Gamma$, $\{\hat{q}_j\}$ be a sequence of characteristic functions having no zeros in some fixed neighborhood of the origin, and $\{\alpha_j\}$ be a sequence of numbers satisfying (9.2.2). If relation (9.2.3) holds for $t = \lambda_{k,l} f_k$ ($k = 1, \ldots, n$; $l = 1, 2, \ldots$) where $\lambda_{k,l}$ are n sequences of decreasing positive numbers converging to zero when $l \to +\infty$ and $\{f_1, \ldots, f_n\}$ is a basis of R^n, then $\hat{q}_j$ is regular in $R^n + i\Gamma$, and relation (9.2.3) holds in any domain of $R^n + i\Gamma$ containing the origin where $\hat{p}$ does not vanish.

Proof For the case $n = 1$, see Ramachandran (1967, Theorem 7.5). In the general case, we define α_k and β_k by

$$\beta_k = \sup\{\lambda \in R : \lambda f_k \in \Gamma\}, \qquad \alpha_k = -\inf\{\lambda \in R : \lambda f_k \in \Gamma\}$$

and we obtain from Corollary 3.3.3 and the case $n = 1$ that $\mathrm{pr}_{f_k} \hat{p}_j$ ($k = 1, \ldots, n$; $j = 1, 2, \ldots$) is regular in $A_k = \{t \in C^n : -\alpha_k < \mathrm{pr}_{f_k} \operatorname{Im} t < \beta_k\}$ and that the relation

$$\mathrm{pr}_{e_k} \hat{p}(t) = \prod_{j=1}^{\infty} [\mathrm{pr}_{e_k} \hat{q}_j(t)]^{\alpha_j}$$

holds in any domain of A_k where $\mathrm{pr}_{f_k} \hat{p}$ does not vanish. From Corollary 3.3.4, it follows that $\hat{p}_j$ $(j = 1, 2, \ldots)$ is regular in the convex hull of the A_k and relation (9.2.3) holds in some neighborhood of the origin. Therefore Theorem 9.2.2 implies Theorem 9.2.3.

9.3 PROBABILITIES WITHOUT INDECOMPOSABLE α-FACTORS

We study in this section the class which we denote by I_α^n of all the n-variate probabilities without indecomposable α-factors (or of all the probabilities which have only infinitely divisible α-factors). We have evidently $I_\alpha^n \subset I_0^n$, but we do not know if I_α^n is really distinct from I_0^n.

If we look at the properties of α-factors, they have some of the properties of factors; for example, we have for α-factors of infinitely divisible probabilities with entire characteristic functions the ridge property. Therefore the proofs of the membership in I_0^n of probabilities based on the ridge property can be extended word for word to the case of membership in I_α^n. For example, we have the following results.

Theorem 9.3.1 Normal and Poisson probabilities belong to I_α^n. More generally, convolutions of a normal and a Poisson probability belong to I_α^n.

Theorem 9.3.2 The probabilities described by Theorems 6.4.3 and 6.4.4 belong to I_α^n.

We have also the relation between supports of purely atomic probabilities given by Theorem 9.1.1. With this relation, we can rewrite for I_α^n the proofs of the membership in I_0^n of most of the purely atomic probabilities. For example, we have

Theorem 9.3.3 Probabilities described by Theorems 8.1.3 and 8.2.6 belong to I_α^n.

We can extend also the isomorphism method. If $A \subset R^n$ and $A' \subset R^{n'}$ are two enumerable independent sets and if ϕ is a bijection of A on A', then the isomorphism Φ induced by ϕ on $\mathscr{M}((Z)A)$ maps α-factors of p on α-factors of $\Phi(p)$. Therefore $\Phi(p)$ belongs to $I_\alpha^{n'}$ if and only if $p \in I_\alpha^n$. With this method, we can prove easily the following results.

Theorem 9.3.4 If p is purely atomic and satisfies conditions of Corollary 7.4.1, p belongs to I_α^n.

Theorem 9.3.5 The probabilities described by Theorems 8.3.1 and 8.3.2 belong to I_α^n.

A particular case is

Theorem 9.3.6 If the characteristic function $\hat{p}$ admits an $(\alpha, 0, \mu)$-De Finetti representation with a Poisson measure concentrated on an enumerable independent set, then p belongs to I_α^n.

With the proofs of Corollaries 7.4.2 and 7.4.3, we deduce from this result

Corollary 9.3.1 The set I_α^n is dense in the set I^n.

Corollary 9.3.2 (a) If Γ is a convex set containing the origin, there exists a probability which belongs to $I_\alpha^n \cap \mathscr{A}_\Gamma$, but does not belong to $\mathscr{A}_{\Gamma'}$ for any convex set Γ' greater than Γ.

(b) There exists a probability which belongs to I_α^n and does not belong to $\mathscr{A}_\Gamma$ for any convex set Γ containing the origin.

Finally, we note that all the preceding results can be extended to enumerable α-decompositions if we assume always condition (9.3.2) for the sequence of components.

NOTES

The first result on α-decompositions is due to Linnik (1956) who proved that univariate normal probabilities belong to I_α^1. Then Dugué (1957) proved that univariate Poisson probabilities belong to I_α^1. Laha (1958) gives the result on univariate analytic characteristic functions and Laha and Lukacs (1962) give some results on derivable characteristic functions. An important subclass of I_α^1 has been described by Linnik (1959c).

In answer to a question raised by Linnik (1955), Mamaĭ (1960) extended Linnik's result on univariate normal probabilities to enumerable α-decompositions when condition (9.2.2) is satisfied. The corresponding result for univariate Poisson probabilities is due to Ramachandran (1964) and Laha (1969). Theorems 9.2.2 and 9.2.3 are due to Ramachandran (1965, 1967) [see also Cuppens (1963b)] in the case of univariate probabilities. Some results on the enumerable α-factors of probabilities with derivable characteristic functions have been given by Cuppens (1963a). The extensions to multivariate probabilities are essentially due to Cuppens (1967). The example of Section 9.1 is due to Dugué (1951b).

Appendix A

Some Results of Function Theory

A.1 STONE–WEIERSTRASS THEOREM

We recall that a set $\mathscr{A}$ of functions defined on E with values in F separates the points of E if for any $(x, y) \in E^2$ $(x \neq y)$, there exists some $f \in \mathscr{A}$ such that $f(x) \neq f(y)$.

We have then

Theorem A.1.1 (*Complex form of the Stone–Weierstrass theorem*) If E is a compact metric space and if a subalgebra $\mathscr{A}$ of $\mathscr{C}_C(E)$ contains the constants, separates the points of E, and is such that, for any $f \in \mathscr{A}$, $\bar{f} \in \mathscr{A}$, then $\mathscr{A}$ is dense (for the uniform metric) *in* $\mathscr{C}_C(E)$.

Proof See Dieudonné (1969, Section 7.3).

From this theorem, we deduce easily

Corollary A.1.1 Let f be a function defined on R^n which is periodic with period T_j with respect to the jth variable $(j = 1, \ldots, n)$. Then f is the uniform limit of a sequence of trigonometric polynomials

$$\sum_{k_1=-N}^{N} \cdots \sum_{k_n=-N}^{N} c_{k_1, \ldots, k_n} \exp\left(2\pi i \sum_{j=1}^{n} k_j t_j / T_j\right).$$

A.2 ALMOST PERIODIC FUNCTIONS

A continuous function f defined on R^n with complex values is an almost periodic function if for any $\varepsilon > 0$ there exists some positive constant l (depending on f and ε) such that any cube $C_a = \{x = (x_1, \ldots, x_n) \in R^n : |x_j - a_j| < l/2, j = 1, \ldots, n\}$ of side l [$a = (a_1, \ldots, a_n \in R^n$] contains an ε-almost period of f, that is, a $\tau \in C_a$ satisfying

$$|f(t + \tau) - f(t)| < \varepsilon$$

for any $t \in R^n$.

Here, we recall only the results which are useful for the theory of Fourier–Stieltjes transforms of finite signed measures.

Theorem A.2.1 (*Favard* (1933, p. 23)) (a) A function which is periodic with respect to each variable is an almost periodic function.

(b) If f_j is an almost periodic function ($j = 1, 2$), then $f_1 + f_2$ and $f_1 f_2$ are almost periodic functions.

(c) If f_j is an almost periodic function ($j = 1, 2, \ldots$) and if the sequence $\{f_j\}$ converges uniformly to f, then f is an almost periodic function.

Corollary A.2.1 If $\{a_j\}$ and $\{l_j\}$ are two sequences ($a_j \in R$, $l_j \in R^n$) such that the series having $a_j e^{i(l_j, t)}$ for general term converges uniformly, then the function f defined by

$$f(t) = \sum_{j=0}^{\infty} a_j e^{i(l_j, t)}$$

is an almost periodic function.

Theorem A.2.2 (*Favard* (1933, p. 24)) If f is an almost periodic function, then

$$a_l = \lim_{T_1 \to \infty} \cdots \lim_{T_n \to \infty} \left(\prod_{j=1}^{n} T_j \right)^{-1} \int_0^{T_1} \cdots \int_0^{T_n} f(t) e^{-i(l, t)} \lambda(dt) \qquad \text{(A.2.1)}$$

exists for any $l \in R^n$ and $\Lambda_f = \{l \in R^n : a_l \neq 0\}$ is countable.

From this theorem, we deduce that an arbitrary almost periodic function f has a Fourier series

$$\sum_{l \in \Lambda_f} a_l e^{i(l, t)}$$

where the Fourier coefficients a_l are defined by (A.2.1).

Theorem A.2.3 (*Favard* (1933, p. 65)) If f and g are two almost periodic functions having the same Fourier series, then $f = g$.

From this result, we deduce easily

Corollary A.2.2 If f is an almost periodic function and if the family $(a_l)_{l \in \Lambda_f}$ of its Fourier coefficients is summable, then

$$f(t) = \sum_{l \in \Lambda_f} a_l e^{i(l, t)}$$

for any $t \in R^n$.

Theorem A.2.4 (*Favard* (1933, p. 81)) Let f and g be two almost periodic functions. If, for any $\varepsilon > 0$, there exists some $\delta(\varepsilon) > 0$ such that any δ-almost period of f is an ε-almost period of g, then

$$(Z)\Lambda_g \subset (Z)\Lambda_f .$$

Theorem A.2.5 Let f be an almost periodic function having a summable family of Fourier coefficients. If F is an analytic function regular on $\overline{f(R)}$, then $g = F \circ f$ is an almost periodic function having a summable family of Fourier coefficients. Moreover $(Z)\Lambda_g \subset (Z)\Lambda_f$.

For the first assertion, see Gelfand *et al.* (1964, p. 175).

We introduce now the notion of analytic almost periodic function. An analytic function f regular in the strip $S = \{t \in C : \sigma_1 < \text{Im}\, t < \sigma_2\}$ is an analytic almost periodic function if for any $\varepsilon > 0$ there exists some positive constant l (depending on ε and f) such that any real interval with length l contains an ε-almost period τ of f, that is, a number τ satisfying

$$| f(t + \tau) - f(t) | < \varepsilon$$

for any $t \in S$.

Theorem A.2.6 (*Favard* (1933, p. 133)) If f is an analytic almost periodic function which is regular and without zeros in $\{t \in C : \sigma_1 < \text{Im}\, t < \sigma_2\}$, then

$$\inf_{\sigma' \leq \text{Im}\, t \leq \sigma''} | f(t) | > 0$$

for any $\sigma_1 < \sigma' \leq \sigma'' < \sigma_2$.

A function u defined on R^2 with real values harmonic in the strip $S = \{(x, y) \in R^2 : \sigma_1 < y < \sigma_2\}$ is an harmonic almost periodic function if for any $\varepsilon > 0$ there exists some positive constant l (depending on ε and u) such that any real interval of length l contains an ε-almost period of u, that is, a number τ satisfying

$$| u(x + \tau, y) - u(x, y) | < \varepsilon$$

for any $(x, y) \in S$.

Theorem A.2.7 (*Favard* (1933, p. 148)) An harmonic almost periodic function u admits a Fourier series

$$ay + b + \sum_{l \in \Lambda_u} (K_l e^{-ly} \cos(lx) + L_l e^{-ly} \sin(lx))$$

where a, b, K_l, and L_l are real constants and Λ_u is a countable set. K_l and L_l are given by

$$\lim_{T \to \infty} (1/T) \int_0^T u(t, y) \cos(lt)\, dt = \tfrac{1}{2}(K_l e^{-ly} + K_{-l} e^{ly}),$$
$$\lim_{T \to \infty} (1/T) \int_0^T u(t, y) \sin(lt)\, dt = \tfrac{1}{2}(L_l e^{-ly} + L_{-l} e^{ly}). \tag{A.2.2}$$

In particular, if $u(-x, y) = u(x, y)$ for any $(x, y) \in S$, then $L_l = 0$ for any $l \in \Lambda_u$.

Since two distinct harmonic almost periodic functions have distinct Fourier series (Favard, 1933, p. 149); from formulas (A.2.2), we deduce easily

Theorem A.2.8 If $((|K_l| + |L_l|)e^{-ly})_{l \in \Lambda_u}$ is a summable family for any $\sigma_1 < y < \sigma_2$, then

$$u(x, y) = ay + b + \sum_{l \in \Lambda_u} (K_l e^{-ly} \cos(lx) + L_l e^{-ly} \sin(lx))$$

for any $(x, y) \in S$.

A.3 INDEPENDENT SETS

A set $A \subset R^n$ is an independent set if it is independent with respect to the rationals, that is, if

$$\sum_{j=1}^{m} l_j a_j = 0$$

($m \in N$, $l_j \in Q$, $a_j \in A$) implies

$$l_1 = \cdots = l_m = 0.$$

The existence of uncountable independent sets is given by

Theorem A.3.1 If a and b are two real numbers, $a < b$, then $[a, b]$ contains a perfect independent set.

Proof See Kahane and Salem (1963, pp. 18–20).

Corollary A.3.1 If $A \subset R^n$ is a Borel independent set and a and b are two real numbers, $a < b$, then $[a, b]$ contains an independent set $A' \in F_\sigma^1$ which is equivalent to A.

Proof This follows immediately from Theorem A.3.1 and the fact that any Borel subset of R^n either is countable or contains a perfect set [cf. Kuratowski (1952, p. 355)].

Now let $A \subset R^n$ and $A' \subset R^{n'}$ be two equipotent independent Borel sets and ϕ be a bijection of A on A'. We can extend ϕ to a bijection Φ of $(N)A$ on $(N)A'$ by the formula

$$\Phi\left(\sum_{j=1}^{m} l_j a_j\right) = \sum_{j=1}^{m} l_j \phi(a_j) \tag{A.3.1}$$

for any $m \in N$, $l_j \in N$, $a_j \in A$.

We recall that a bijection f of a Borel set C on a Borel set C' is bimeasurable if $f(B)$ is a Borel subset of C' for any Borel subset B of C and $f^{-1}(B')$ is a Borel subset of C for any Borel subset B' of C'.

We have then

Theorem A.3.2 If $A \in F_\sigma^n$ and $A' \in F_\sigma^{n'}$ are two equipotent independent sets and if ϕ is a bimeasurable bijection of A on A', then the mapping Φ defined by (A.3.1) is a bimeasurable bijection of $(N)A$ on $(N)A'$.

Proof We will use the following theorem which follows directly from a theorem of Lusin (1930, p. 178):

Let E be an m-dimensional subspace of R^n and $B \in \mathscr{B}_n$. If $B \cap \mathrm{pr}_E^{-1}(x)$ is countable for any $x \in E$, then $\mathrm{pr}_E\, B$ is a Borel subset of E.

Now let $m \in N$ and Φ_m be the restriction of Φ to $(m)A$. Since

$$(m_1)A \cap (m_2)A = \varnothing, \qquad (m_1)A' \cap (m_2)A' = \varnothing \tag{A.3.2}$$

for any $m_1 \neq m_2$, Φ_m is a bijection of $(m)A$ on $(m)A'$. If we consider $A^m = A \times \cdots \times A$ (m times) and $A'^m = A' \times \cdots \times A'$ (m times). $A^m \in F_\sigma^{mn}$ and $A'^m \in F_\sigma^{mn'}$ and

$$\Phi_m = P'_m \circ \phi_m \circ P_m^{-1} \tag{A.3.3}$$

where ϕ_m is the mapping of A^m on A'^m defined by

$$\phi_m(a_1, \ldots, a_m) = (\phi(a_1), \ldots, \phi(a_m))$$

($a_j \in A$, $j = 1, \ldots, m$), P_m is the mapping of A^m on $(m)A$ defined by

$$P_m(a_1, \ldots, a_m) = a_1 + \cdots + a_m$$

($a_j \in A$, $j = 1, \ldots, m$), and P'_m is the mapping of A'^m on $(m)A'$ defined by

$$P'_m(a'_1, \ldots, a'_m) = a'_1 + \cdots + a'_m$$

($a'_j \in A'$, $j = 1, \ldots, m$).

If $\mathscr{C}$ is the class of all the Borel subsets B_m of A^m such that $\phi_m(B_m)$ is a Borel subset of A'_m, then $\mathscr{C}$ is a σ-ring containing all the sets $C_1 \times \cdots \times C_m$ where

C_j is a Borel subset of A. Therefore $\mathscr{C}$ is the class of all the Borel subsets of A^m. Since we can do the same for ϕ_m^{-1}, ϕ_m is a bimeasurable bijection of A^m on A'^m. Moreover, since P_m is continuous, $P_m^{-1}(B)$ is a Borel subset of A^m for any Borel subset B of $(m)A$.

Finally, P'_m is the composition of the projection of $(m)A$ on the "diagonal subspace" $\{x = (x_1, \ldots, x_m) : x_j \in R^n, x_1 = \cdots = x_m\}$ with a canonical mapping of this diagonal subspace on R^n. Since, for any $x' \in (m)A'$, $P_m'^{-1}(x')$ is finite, we obtain from Lusin's theorem that $P'_m(B'_m)$ is a Borel subset of $(m)A'$ for any Borel subset B'_m of A'^m. It follows then from (A.3.3) that $\Phi_m(B)$ is a Borel subset of $(m)A'$ for any Borel subset B of $(m)A$ and, since we can do the same for Φ_m^{-1}, Φ_m is a bimeasurable bijection of $(m)A$ on $(m)A'$.

Now let B be a Borel subset of $(N)A$ and B' be a Borel subset of $(N)A'$. From (A.3.2), we obtain that

$$\Phi(B) = \bigcup_{m=0}^{\infty} \Phi_m(B_m)$$

where $B_m = B \cap (m)A$ is a Borel subset of $(m)A$ $(m = 0, 1, 2, \ldots)$ and

$$\Phi^{-1}(B') = \bigcup_{m=0}^{\infty} \Phi_m^{-1}(B'_m)$$

where $B'_m = B' \cap (m)A'$ is a Borel subset of $(m)A'$ $(m = 0, 1, 2, \ldots)$. Since Φ_m is bimeasurable $(m = 0, 1, 2, \ldots)$, we obtain easily the bimeasurability of Φ.

From the preceding theorem, we must study the bimeasurable bijections between two independent Borel sets, but we have

Theorem A.3.3 If $A \subset R^n$ and $A' \subset R^{n'}$ are two equipotent Borel sets, there exists a bimeasurable bijection of A on A'.

Proof See Kuratowski (1952, p. 358).

From this theorem and from Corollary A.3.1, we have

Theorem A.3.4 If $A \subset R^n$ is an independent Borel set and a and b are two real numbers, $a < b$, there exists a bimeasurable bijection of A on an independent subset $A' \in F_\sigma^1$ of $[a, b]$.

We use some extensions of the preceding result.

Theorem A.3.5 If $A = \bigcup_{j=1}^{\infty} A_j \subset R^n$ is an independent Borel set $(A_j \cap A_k = \varnothing$ if $j \neq k)$ and $\{a_j\}$ and $\{b_j\}$ are two sequences of real numbers, $a_j < b_j$ $(j = 1, 2, \ldots)$, there exists a bimeasurable bijection ϕ of A on an independent set $A' \in F_\sigma^1$ such that

$$\phi(A_j) \subset]a_j, b_j[$$

$(j = 1, 2, \ldots)$.

Proof We can suppose without loss of generality that $a_j b_j > 0$ $(j = 1, 2, \ldots)$. From the preceding theorem, there exists a bimeasurable bijection ψ of A on an independent subset $C \in F_\sigma^1$ of $[1, 2]$. If $\psi(A_j) = C_j$, we can choose some $\alpha_j \in C_j$ and some $n_j \in N$ such that

$$(2 + n_j\alpha_j)(1 + n_j\alpha_j)^{-1} < \begin{cases} b_j a_j^{-1} & \text{if } a_j > 0, \\ a_j b_j^{-1} & \text{if } a_j < 0, \end{cases}$$

then some $m_j \in Q$ such that

$$a_j(1 + n_j\alpha_j)^{-1} < m_j < b_j(2 + n_j\alpha_j)^{-1}$$

if $a_j > 0$ and

$$a_j(2 + n_j\alpha_j)^{-1} < m_j < b_j(1 + n_j\alpha_j)^{-1}$$

if $a_j < 0$.

If we consider the function θ_j defined on C_j by

$$\theta_j(x) = m_j(x + n_j\alpha_j),$$

we have a bimeasurable bijection of C_j on $\theta_j(C_j) = A'_j$. Let now ϕ be the mapping defined on A by

$$\phi(x) = \theta_j(\psi(x))$$

if $x \in A_j$. Then $\phi(A) = A'$ is an independent set and belongs to F_σ^1 and $\phi(A_j) \subset]a_j, b_j[$. Now if B is a Borel subset of A, then

$$\phi(B) = \bigcup_{j=1}^{\infty} \phi(B_j)$$

where $B_j = B \cap A_j$, and since ψ and θ_j are bimeasurable bijections, we obtain that $\phi(B)$ is a real Borel set. By the same way, if B' is a Borel subset of A', then $\phi^{-1}(B')$ is a Borel set and the theorem is proved.

Theorem A.3.6 Let $A = \bigcup_{j=1}^{p} A_j$ be an independent Borel set of R^n $(A_j \cap A_k = \varnothing$ if $j \neq k)$ and $\{\theta_1, \ldots, \theta_p\}$ be a basis of R^p. If ϕ is a bimeasurable bijection of A on an independent set $A' \in F_\sigma^1$, then the mapping ψ defined by

$$\psi(x) = \phi(x)\theta_j$$

if $x \in A_j$ is a bimeasurable bijection of A on an independent subset $A' \in F_\sigma^p$.

Proof It is clear that $\psi(A)$ belongs to F_σ^p and is an independent set and that ψ is a bijection of A on $\psi(A)$. Now, if B is a Borel subset of A_j, then ψ is the composition of two bimeasurable bijections and $\psi(B)$ is a Borel subset of R^p. If B is a Borel subset of A, then

$$\psi(B) = \bigcup_{j=1}^{p} \psi(B_j)$$

where $B_j = B \cap A_j$ and $\psi(B)$ is a Borel subset of R^p. Likewise, for any Borel subset B' of $\psi(A)$, $\psi^{-1}(B')$ is a Borel subset of A and the theorem is proved.

A.4 ANALYTIC FUNCTIONS

In this book, we use many simple results on analytic and entire functions: for example, Titchmarsh (1939) for functions of one complex variable and Fuks (1963) for functions of several complex variables. Here we give only some special results.

Theorem A.4.1 Let f be a univariate function regular in a domain containing the circle $\{\xi : |\xi - z_0| \le R\}$ and u and v be the real and the imaginary part of f. Then,

$$f(z) = \frac{1}{2\pi}\int_0^{2\pi} u(x_0 + \rho\cos\theta, y_0 + \rho\sin\theta)\frac{\rho e^{i\theta} + z - z_0}{\rho e^{i\theta} - z + z_0}\,d\theta + iy(z_0)$$

for any z satisfying $|z - z_0| < \rho < R$ $(z_0 = x_0 + iy_0,\ x_0 \in R,\ y_0 \in R)$ and

$$f'(z_0) = \frac{1}{\pi\rho}\int_0^{2\pi} u(x_0 + \rho\cos\theta, y_0 + \rho\sin\theta)e^{-i\theta}\,d\theta.$$

The first formula is a formulation of the well-known Poisson's formula [see Titchmarsh (1939, p. 125)]. If we differentiate this formula with respect to z and put $z = z_0$, we obtain the second one.

Corollary A.4.1 Let f be an n-variate entire function and u be the real part of f. If

$$u(x, y) = O(\exp(\tau + \varepsilon)\|z\|)$$

when $\|z\| \to +\infty$ for any $\varepsilon > 0$, then

$$f(z) = O(\exp(\tau + 3\varepsilon)\|z\|)$$

when $\|z\| \to +\infty$.

Proof Let $\theta \in R^n$, $\theta \ne 0$, $\lambda = \mu + iv \in C$, $\phi(\lambda) = f(\lambda\theta)$, and $u(\mu\theta, v\theta) = \omega(\mu, v)$. Then ϕ and ω are related by the Poisson formula

$$\phi(\lambda) = i\,\mathrm{Im}\,\phi(0) + \frac{1}{2\pi}\int_0^{2\pi} \omega(\rho\cos\theta, \rho\sin\theta)\frac{\rho e^{i\theta} + \lambda}{\rho e^{i\theta} - \lambda}\,d\theta$$

if $|\lambda| < \rho$. Letting $\rho = |\lambda| + \varepsilon$ and $|\lambda| \to +\infty$, we obtain

$$|\phi(\lambda)| = O(\exp(\tau + 3\varepsilon)\|\lambda\theta\|)$$

and since θ is arbitrary, we obtain the corollary.

Theorem A.4.2 (*Phragmén–Lindelöf theorem*) Let f be a function of one complex variable regular in $D = \{z : |\arg z| < \pi/2\alpha\}$, continuous in $\bar{D}$, and satisfying the following conditions:

(a) $f(z) = O(\exp|z|^\beta)$ when $|z| \to +\infty$ for some $\beta < \alpha$;
(b) $|f(z)| \leq M$ if $z \in \partial D$.

Then $|f(z)| \leq M$ if $z \in D$.

Proof See Titchmarsh (1939, p. 177).

We introduce now the following definitions.

A domain $D \subset C^n$ is the *domain of analyticicity of* f if f is regular in D and f cannot be continued beyond the boundary ∂D of D.

If $\mathscr{F}$ is a class of functions which are regular in D, a point $\xi \in \partial D$ has the *frontier property* for the class $\mathscr{F}$ if, for any compact set $\Delta \subset D$ and any $\varepsilon > 0$, there exist some function $f \in \mathscr{F}$ and some point $z \in D\backslash\Delta$ such that

$$\|z - \xi\| < \varepsilon \qquad \text{and} \qquad |f(t)| \leq 1$$

for $t \in \Delta$ while

$$|f(z)| > 1.$$

Theorem A.4.3 (*Cartan–Thullen theorem*) Let D be a domain of C^n and $\mathscr{F}$ be a class of functions regular in D which is closed for addition, multiplication by a positive constant, raising to a positive integer power, and passing to the limit uniformly on each compact set of D. If each point of ∂D has the frontier property for the class $\mathscr{F}$, there exists some function $f \in \mathscr{F}$ having D for domain of analyticity.

Proof There exists a countable set Γ of points of ∂D which is dense on D. Let $\{\xi_m\}$ be a sequence of points of Γ such that each point of Γ occurs in $\{\xi_m\}$ infinitely often and let $\{\Gamma_m\}$ be a sequence of compact sets in D such that $\Gamma_m \subset \Gamma_{m+1}$, $\bigcup_m \Gamma_m = D$. Since ξ_1 has the frontier property for $\mathscr{F}$, there exist $f_1 \in \mathscr{F}$ and $z_1 \notin \Gamma_1$ such that $\|z_1 - \xi_1\| < 1$ and $|f_1(t)| \leq 1$ for $t \in \Gamma_1$ while $|f_1(z_1)| > 1$. Now there exists some integer r_1 such that $\Delta_2 = \Gamma_{r_1} \supset \Delta_1 \cup \{z_1\}$ and, since ξ_2 has the frontier property for $\mathscr{F}$, there exist $f_2 \in \mathscr{F}$ and $z_2 \in D\backslash\Delta_2$ satisfying $\|z_2 - \xi_2\| < \frac{1}{2}$ and $|f_2(t)| \leq 1$ for $t \in \Delta_2$ while $|f_2(z_2)| > 1$ and so on. Step by step, we construct a sequence $\{\Delta_m\}$ of compact sets of D, a sequence $\{z_m\}$ of points of D and a sequence $\{f_m\}$ of functions of $\mathscr{F}$ satisfying the following conditions:

(a) $z_m \in \Delta_{m+1}\backslash\Delta_m$;
(b) $\|z_m - \xi_m\| < 1/m$;
(c) $|f_m(t)| \leq 1$ for $t \in \Delta_m$ while $|f_m(z_m)| > 1$.

From (c), we deduce easily the existence of some integer s_m (with $s_1 = 1$) such that

$$\frac{|f_m(z_m)|^{s_m}}{m^2} - \sum_{j=1}^{m-1} \frac{|f_j(z_m)|^{s_j}}{j^2} > m.$$

We define f by

$$f = \sum_{j=1}^{\infty} \frac{(f_j(z))^{s_j}}{j^2}.$$

Since $|f_j(z)| \leq 1$ for $z \in \Delta_m$ and $j < m$, the series converges uniformly on each Δ_m and $f \in \mathscr{F}$. From

$$|f(z_m)| \geq \frac{|f_m(z_m)|^{s_m}}{m^2} - \sum_{j=1}^{m-1} \frac{|f_j(z_m)|^{s_j}}{j^2} - \sum_{j=m+1}^{\infty} \frac{1}{j^2}$$

we deduce that

$$\lim_{j\to\infty} |f(z_j)| = +\infty$$

and D is the domain of analyticity of f.

A.5 TOPOLOGICALLY INDEPENDENT FUNCTIONS

Let a and b be two real numbers, $a < b$, and $\mathscr{C}(a, b)$ be the linear space of all the continuous functions defined on $[a, b]$ with complex values. This linear space is normed with the usual norm

$$\|f\| = \sup_{a\leq x\leq b} |f(x)|$$

if $f \in \mathscr{C}(a, b)$.

A sequence $\{f_j\}_{j=-\infty}^{+\infty}$ ($f_j \in \mathscr{C}(a, b)$) is topologically independent in $\mathscr{C}(a, b)$ if

$$\lim_k \left(\sum_{j=-\infty}^{+\infty} a_{j,k} f_j \right) = 0$$

($a_{j,k} \in C$, the limit being taken with respect to the norm) implies

$$\lim_k a_{j,k} = 0$$

($j = 0, \pm 1, \pm 2, \ldots$). This condition is equivalent to the following one: If

$$g = \lim_k \left(\sum_{j=-\infty}^{+\infty} a_{j,k} f_j \right)$$

($g \in C(a, b)$, $a_{j,k} \in C$), $\lim_k a_{j,k}$ exists for $j = 0, \pm 1, \pm 2, \ldots$.

Theorem A.5.1 In $\mathscr{C}(a, b)$, a sequence $\{f_j\}$ is topologically independent if and only if there exists some measure μ concentrated on $[a, b]$ and orthogonal to f_j for any j, that is, a measure μ satisfying

$$\int f_j(x)\mu(dx) = 0$$

for any j.

Proof See Schwartz (1959, pp. 17–19).

We prove now

Theorem A.5.2 Let $\{\lambda_j\}_{j=-\infty}^{+\infty}$ be a sequence of real numbers satisfying

$$\cdots < \lambda_{-2} < \lambda_{-1} < \lambda_0 = 0 < \lambda_1 < \lambda_2 < \cdots$$

and

$$\sum_{j=-\infty}^{+\infty} |\lambda_j|^{-1} < +\infty.$$

The sequence $\{z, z^2, \exp(\lambda_j z)\}$ is topologically independent in $\mathscr{C}(a, b)$.

Proof Let ϕ be the function defined by

$$\phi(t) = t^3 \prod_{j=-\infty}^{+\infty} \left| \left(1 - \frac{it}{\lambda_j}\right) \frac{\sin^2(\alpha_j t)}{\alpha_j^2 t^2} \right|$$

where $\{\alpha_j\}$ is a sequence of positive numbers satisfying

$$\alpha = \sum_{j=-\infty}^{+\infty} \alpha_j < +\infty.$$

Then ϕ satisfies the following conditions:

(a) ϕ is an entire function of exponential type 2α;
(b) $\phi \in L^2(R)$;
(c) $\phi(0) = \phi'(0) = \phi''(0) = 0$;
(d) $\phi(-i\lambda_j) = 0$ $(j = \pm 1, \pm 2, \ldots)$.

We deduce then from Paley–Wiener's theorem (cf. Theorem 3.6.2) the existence of an absolutely continuous signed measure μ such that

$$\phi(t) = \int_{-2\alpha}^{2\alpha} e^{itx}\mu(dx).$$

If $4\alpha = b - a$, we obtain

$$\phi(t) \exp\left(i\left(\frac{a+b}{2}\right)t\right) = \int_a^b e^{itx}\mu'(dx)$$

where $\mu' = \mu * \delta_{(a+b)/2}$ is concentrated on $[a, b]$ and is orthogonal to the functions z, z^2, and $\exp(\lambda_j z)$ ($j = 0, \pm 1, \pm 2, \ldots$) and this, with the preceding result, implies the theorem.

We use also the following result which is a particular case of a theorem of Schwartz (1959, pp. 57–58).

Theorem A.5.3 Let $\{\lambda_j\}$ be a sequence of real numbers satisfying the following conditions:

(a) $\cdots < \lambda_{-2} < \lambda_{-1} < \lambda_0 = 0 < \lambda_1 < \lambda_2 < \cdots$;
(b) $\sum_{j=-\infty}^{+\infty} |\lambda_j|^{-1} < +\infty$;
(c) $\lambda_{j+1} - \lambda_j \geq q > 0$.

Also $f_j(z) = \exp(\lambda_j z)$. If

$$g = \lim_k \sum_{j=-\infty}^{+\infty} a_{j,k} f_j,$$

then g is regular in the strip $\{z \in C : a < \operatorname{Re} z < b\}$ and

$$g(z) = \sum_{j=-\infty}^{+\infty} \left(\lim_k a_{j,k}\right) f_j(z),$$

the convergence being normal in any compact of this strip.

Appendix B

Exponentials of Polynomials and Functions

In this appendix, we prove the results stated in Section 8.4 on exponentials of polynomials and functions. These results are due to Lévy (1937a, 1938b) for the case $n = 1$ and to Cuppens (1970) for the general case.

B.1 CASE OF A POLYNOMIAL OF ONE VARIABLE

Let P be the polynomial

$$P(x) = \sum_{j=1}^{p} a_j x^j \tag{B.1.1}$$

and F be the power series

$$F(x) = \exp(P(x)) = \sum_{k=0}^{\infty} A_k x^k. \tag{B.1.2}$$

By derivation, we obtain

$$F'(x) = P'(x)F(x)$$

and, replacing F, F', and P' by their development, we obtain the inductive equation

$$(m + p)A_{m+p} = \sum_{j=0}^{p} ja_j A_{m+p-j} \cdot \tag{B.1.3}$$

Letting

$$B_k = A_k \Gamma((k/p) + 1)$$

and assuming that $a_p = 1$ (this does not restrict the generality), we obtain from (B.1.3)

$$B_{m+p} - B_m = \sum_{j=1}^{p-1} \phi_j(m) a_{p-j} B_{m+j} \tag{B.1.4}$$

where

$$\phi_j(m) = \frac{p-j}{p} \frac{\Gamma((m/p) + 1)}{\Gamma((m + j/p) + 1)}.$$

From the linearity of Eq. (B.1.4), we deduce that if $B_{k,m}$ $(k = 1, \ldots, p)$ are p linearly independent solutions of (B.1.4), then

$$B_m = \sum_{k=1}^{p} l_k B_{k,m}$$

is the general solution of (B.1.4), l_k being some constants.

We will search independent solutions of (B.1.4) by using the theory of inductive equations with constant coefficients. For fixed m, (B.1.4) is such an equation and we search solutions of the kind

$$B_{m+j} = q^j(m)$$

$(j = 0, \ldots, p)$. We have then

$$q^p(m) - 1 = \sum_{j=1}^{p-1} \phi_j(m) a_{p-j} q^j(m). \tag{B.1.5}$$

From Stirling's formula, we deduce immediately that

$$\lim_{m \to \infty} m^{j/p} \phi_j(m) = l > 0, \tag{B.1.6}$$

so that

$$\lim_{m \to \infty} \phi_j(m) = 0.$$

This implies that (B.1.5) has p roots $q_k(m)$ tending, when $m \to +\infty$, to

$$r_k = \exp(2ik\pi p^{-1})$$

$(k = 0, \ldots, p-1)$. If m is great enough $(m \geq M_0)$, the roots $q_k(m)$ are all distinct (we denote by q_k the root which is the nearest from r_k) and $q_0(m)$ is positive. The solution of (B.1.4) is, for $m \geq M_0$,

$$B_{m+j} = \sum_{k=0}^{p-1} c_k(m) q_k^j(m) \tag{B.1.7}$$

$(j = 0, 1, \ldots, p)$, the $c_k(m)$ being independent of j.

Knowing B_{m+j} ($j = 0, \ldots, p-1$), we can use (B.1.7) for determining the $c_k(m)$ in function of these B_j, the determinant being in this case

$$D_m = \begin{vmatrix} 1 & \cdots & 1 \\ q_0(m) & \cdots & q_{p-1}(m) \\ \vdots & \vdots & \vdots \\ q_0^{p-1}(m) & \cdots & q_{p-1}^{p-1}(m) \end{vmatrix}.$$

When $m \to +\infty$, D_m tends to

$$D = \begin{vmatrix} 1 & \cdots & 1 \\ r_0 & \cdots & r_{p-1} \\ \vdots & \vdots & \vdots \\ r_0^{p-1} & \cdots & r_{p-1}^{p-1} \end{vmatrix} \neq 0$$

and this implies that

$$L_m = \sup_{0 \le k \le p-1} |c_k(m)| \qquad \text{and} \qquad L'_m = \sup_{0 \le k \le p-1} |B_{m+k}|$$

are of the same order when $m \to +\infty$.

We have also

$$B_{m+1+j} = \sum_{k=0}^{p-1} c_k(m+1) q_k^j(m+1)$$

($j = 0, \ldots, p-1$) and from these equations and (B.1.7), we obtain

$$\sum_{k=0}^{p-1} c_k(m) q_k^{j+1}(m) = \sum_{k=0}^{p-1} c_k(m+1) q_k^j(m+1)$$

($j = 0, \ldots, p-1$). Letting

$$Q_k(m) = \prod_{j=M_0}^{m} q_k(j), \qquad c_k(m) = \gamma_k(m) Q_k(m),$$

we obtain

$$\sum_{k=0}^{p-1} (Q_k(m) q_k^{j+1}(m)(\Delta\gamma_k)(m) + c_k(m)(\Delta q_k^{j+1})(m)) = 0$$

($j = 0, \ldots, p-1$) where Δf is defined by

$$\Delta f(m) = f(m+1) - f(m).$$

In these equations, let us consider $Q_k(m)\,\Delta\gamma_k(m)$ as the unknown. The determinant is D_{m+1} which tends to D when $m \to +\infty$, so that $|D_{m+1}|$ is greater than a positive constant when m is great enough ($m \ge M_1$). Since

$$(\Delta q_k^{j+1})(m) = \Delta q_k(m)\left(\sum_{l_1+l_2=j} q_k^{l_1}(m) q_k^{l_2}(m+1)\right),$$

we have

$$|Q_k(m)\,\Delta\gamma_k(m)| \leq K_1 L_m \sup|\Delta q_k(m)| \tag{B.1.8}$$

for $m \geq M_1$ where K_1 is some constant.

If m is great enough, the $q_k(m)$ are holomorphic functions of $\phi_j(m)$ $(j = 1, \ldots, p-1)$ and

$$\Delta q_k(m) = O(\Delta\phi_j(m))$$

when $m \to +\infty$. From the definition of ϕ_j, we prove easily that

$$\Delta\phi_j(m) \sim -(j(p-j)/p^2 m)(p/m)^{j/p}$$

when $m \to +\infty$, so that

$$\Delta q_k(m) = O(1/m^{1+1/p})$$

when $m \to +\infty$ and this, with (B.1.8), implies

$$|Q_k(m)\,\Delta\gamma_k(m)| \leq K_2 L_m/m^{1+1/p} \tag{B.1.9}$$

$(k = 0, \ldots, p-1)$ if $m \geq M_2$ where K_2 is some constant.

Let now η_k be defined by

$$c_k(m) = \eta_k(m) L_m$$

(by definition, one of the $\eta_k(m)$ is equal to 1 and the others lie between 0 and 1) and let ε and η be two constants satisfying $\varepsilon > 0$, $0 < \eta < 1$. If we have

$$\eta_k(m) \geq \eta \tag{B.1.10}$$

for some $m \geq M_2$ and some k, then

$$|\gamma_k(m) Q_k(m)| \geq \eta L_m \tag{B.1.11}$$

and from (B.1.9)

$$|\Delta\gamma_k(m)/\gamma_k(m)| \leq K_2/\eta m^{1+1/p}.$$

Since

$$\Delta \log \gamma_k(m) = \log(1 + \Delta\gamma_k(m)/\gamma_k(m)),$$

$\Delta \log \gamma_k(m)$ is the general term of an absolutely convergent series and this implies that

$$\lim_{m\to\infty} \gamma_k(m) = \gamma_k \neq 0 \tag{B.1.12}$$

if (B.1.10) is satisfied for some k and any m great enough. Moreover, if (B.1.10) is satisfied for m great enough and if

$$\sum_{k=m}^{\infty} K_2/\eta k^{1+1/p} < \varepsilon$$

[this is true if $m \geq M_3(\varepsilon, \eta)$], then the variation of $\log \gamma_k(m)$ is in absolute value less than ε and, up to an error less than $\varepsilon' = e^{\varepsilon} - 1$, $c_k(m)$ is proportional to $Q_k(m)$.

The $q_k(m)$ being holomorphic functions of the $\phi_j(m)$ and the $\phi_j(m)$ being holomorphic functions of $m^{1/p}$, we deduce that

$$q_k(m) = 1 + \sum_{j=1}^{\infty} \alpha_j(k)/m^{j/p}$$

and consequently

$$|q_k(m)| = 1 + \sum_{j=1}^{\infty} \beta_j(k)/m^{j/p}$$

if m is great enough. From this we deduce easily that either $|Q_{k_1}(m)/Q_{k_2}(m)|$ tends to a positive finite limit when $m \to +\infty$ (we say in this case that q_{k_1} and q_{k_2} are roots of the same type) or $|Q_{k_1}(m)/Q_{k_2}(m)|$ tends monotonically (if $m \geq M_4$) to 0 or $+\infty$.

Let us suppose that $q_{k_1}(m)$ and $q_{k_2}(m)$ are not roots of the same type and that (B.1.10) is satisfied for $k = k_j$ ($j = 1, 2$) and any m great enough. From the last remark and the fact that γ_{k_j} has a positive finite limit ($j = 1, 2$), we deduce that $|c_{k_1}(m)/c_{k_2}(m)|$ tends to 0 or $+\infty$. But if (B.1.10) is satisfied, $|c_{k_1}(m)/c_{k_2}(m)|$ lies always between η and $1/\eta$. This implies that inequalities (B.1.10) cannot be satisfied for $k = k_j$ ($j = 1, 2$) and any m great enough.

We study now the $c_k(m)$ which are not $o(L_m)$ when $m \to +\infty$. We suppose that $c_{k_1}(m)$ is not $o(L_m)$ and that $q_{k_1}(m)$ is of the greatest type possible. Since $c_{k_1}(m)$ is not $o(L_m)$, there exists some constant η ($0 < \eta < \frac{1}{2}$) and a sequence of integers m_j such that

$$\eta_{k_1}(m_j) \geq 2\eta.$$

From this we easily deduce that

$$\eta_{k_1}(m) \geq \eta$$

for any m great enough. Let us suppose that $c_{k_2}(m)$ is not $o(L_m)$ when $m \to +\infty$ and that $q_{k_1}(m)$ and $q_{k_2}(m)$ are not of the same type. There exists a sequence of integers m'_j such that

$$\eta_{k_2}(m'_j) < \eta$$

and from

$$\begin{aligned}
\eta_k(m+1) &= \frac{|\gamma_k(m+1)Q_k(m+1)|}{L_{m+1}} \\
&= |q_k(m+1)| \frac{L_m}{L_{m+1}} \frac{|\gamma_k(m+1)Q_k(m)|}{L_m} \\
&\leq |q_k(m+1)| \frac{L_m}{L_{m+1}} \left(\eta_k(m) + \frac{|\Delta\gamma_k(m)Q_k(m)|}{L_m} \right),
\end{aligned}$$

we obtain easily that

$$\eta_k(m'_j + 1) < 3\eta/2.$$

But, if

$$\eta \leq \eta_{k_2}(m) < 3\eta/2,$$

then $\eta_{k_2}(m)$ is the product of $|\gamma_{k_2}(m)|$ which does not much vary and $Q_{k_2}(m)/L_m$ which tends to 0 and therefore

$$\eta_{k_2}(m) < 2\eta$$

for any m great enough. Since η can be chosen arbitrarily, we have

$$c_{k_2}(m) = o(L_m)$$

when $m \to +\infty$ and all the $c_k(m)$ which are not $o(L_m)$ correspond to roots of the same type.

From this and from (B.1.12), we deduce that any solution of Eq. (B.1.4) is of the kind

$$B_m = \sum \gamma_k Q_k(m) + o(Q_k(m)) \tag{B.1.13}$$

when $m \to +\infty$ where the summation is taken on the k such that $q_k(m)$ are of the same type and γ_k are some constants.

Let $q_k(m)$ be a root of (B.1.5) and let h be the number of linearly independent solutions of the kind (B.1.13) corresponding to this root. If there exist h' roots of the same type and if $h > h'$, we can find $h - h'$ linearly independent solutions of (B.1.4) which are $o(Q_k(m))$ when $m \to +\infty$. We order the roots according to the growth of $Q_k(m)$ and begin with the $q_k(m)$ which correspond to the smallest growth of $Q_k(m)$. If there are h roots of this type, we can find at most h linearly independent solutions of the kind (B.1.13) and so on. Since we must have p linearly independent solutions, then to h roots of the same type correspond exactly h linearly independent solutions of the kind (B.1.13) and by linear combination of these one we can find a solution of the kind

$$B_m = Q_k(m) + o(Q_k(m))$$

when $m \to +\infty$. Therefore we have proved

Lemma B.1.1 To each root $q_k(m)$ of Eq. (B.1.5) corresponds some solution $B_{m,k}$ of Eq. (B.1.4) such that

$$B_{m,k} = Q_k(m) + o(Q_k(m))$$

when $m \to +\infty$ and the general solution of (B.1.4) is

$$B_m = \sum_{k=0}^{p-1} l_k B_{m,k}$$

where the l_k are some constants.

Now let P be the polynomial (B.1.1). We can suppose without loss of generality that a_p and the g.c.d. δ of $\{j : a_j \neq 0\}$ are equal to 1. For any integer k, $0 < k < p$, we denote by $\sigma(k)$ the smallest positive integer satisfying

$$a_{p-\sigma(k)} \neq 0, \qquad \cos(2k\sigma(k)\pi/p) < 1.$$

We prove the existence of $\sigma(k)$. Indeed if

$$\cos(2kh\pi/p) = 1$$

for any h satisfying $a_{p-h} \neq 0$, then

$$\cos(2k\delta\pi/p) = 1$$

and this is impossible if $\delta = 1$ and $0 < k < p$.

For any integer j, $0 < j < p$, we introduce the sets

$$\mathfrak{A}_j = \{k : a_k \neq 0, k > j\}, \qquad \mathfrak{A}'_j = \{k : a_k > 0, k > j\},$$

$$\mathfrak{B}_j = \{k : a_k \neq 0, k < j\}, \qquad \mathfrak{B}'_j = \{k : a_k > 0, k < j\}$$

and prove the following.

Lemma B.1.2 The following conditions are equivalent:

- (A_1) for any $a_j < 0$, $j \in (Z)\mathfrak{A}_j$;
- (A_2) for any $a_j < 0$, the g.c.d. δ_j of $\mathfrak{A}_j$ divides j;
- (A_3) for any $a_j < 0$, $j \in (Z)\mathfrak{A}'_j$;
- (A_4) for any $a_j < 0$, the g.c.d. δ'_j of $\mathfrak{A}'_j$ divides j;
- (A_5) $a_{p-\sigma(k)} > 0$ for any $k = 1, 2, \ldots, p-1$.

Proof The equivalence between (A_1) and (A_2) [resp. (A_3) and (A_4)] follows immediately from Bezout's relation.

Evidently (A_3) implies (A_1). Now, if (A_3) is not satisfied, there exists some j such that $j \notin (Z)\mathfrak{A}'_j$. Let m be the greatest of such j. We prove easily that any integer k such that $k > j$, $a_k < 0$ belongs to $(Z)\mathfrak{A}'_m$ and this implies that $(Z)\mathfrak{A}_m = (Z)\mathfrak{A}'_m$ and (A_1) is not satisfied.

We suppose now that (A_5) is not satisfied. There exists some k such that $a_{p-\sigma(k)} < 0$ and let $m = \sup\{p - \sigma(k) : a_{p-\sigma(k)} < 0\}$. Then by definition of

$$\cos(2kj\pi/p) = 1$$

if $j \in \mathfrak{A}_m$ and

$$\cos(2k\delta_m\pi/p) = 1$$

where δ_m is the g.c.d. of $\mathfrak{A}_m$ and since

$$\cos(2km\pi/p) = 1,$$

δ_m does not divide m and (A_2) is not satisfied.

Finally, if we suppose that (A_2) is not satisfied, there exists some j such that $a_j < 0$ and the g.c.d. δ_j of $\mathfrak{A}_j$ does not divide j. Since $p \in \mathfrak{A}_j$, δ_j divides p and it is easily proved that

$$\sigma(p/\delta_j) = p - j,$$

so that

$$a_{p-\sigma(p/\delta_j)} = a_j < 0$$

and (A_5) is not satisfied.

Lemma B.1.3 The following conditions are equivalent:

(B_1) for any $a_j < 0$, $j \in (N)\mathfrak{B}_j$;
(B_2) for any $a_j < 0$, $j \in (N)\mathfrak{B}'_j$.

Proof Evidently (B_2) implies (B_1). If (B_2) is not satisfied, there exists some j such that $a_j < 0$ and $j \notin (N)\mathfrak{B}'_j$. Let m be the smallest of these j. We prove easily that any integer k such that $k < j$, $a_k < 0$ belongs to $(N)\mathfrak{B}'_m$ and this implies that $(N)\mathfrak{B}_m = (N)\mathfrak{B}'_m$ and (B_1) is not satisfied.

We will call "Condition A" one of the conditions of Lemma B.1.2 and "Condition B" one of the conditions of Lemma B.1.3.
We prove now

Theorem B.1.1 Let P be the polynomial (B.1.1). All the coefficients A_k of the expansion (B.1.2) are nonnegative except a finite number if and only if P satisfies Condition A.

Proof *Necessity* If P does not satisfy Condition A, let m be the greatest of the j such that a_j is negative and δ_j does not divide j. If $\rho = \exp(2i\pi\delta_m^{-1})$, then $\rho^k = 1$ if $a_k \neq 0$, $k > m$ and $\rho^m \neq 1$. Therefore $P(x) - P(\rho x)$ is equivalent to $a_m(1 - \rho^m)x^m$ when $x \to +\infty$, so that

$$\mathrm{Re}(P(x) - P(\rho x)) \to -\infty$$

and

$$|F(\rho x)/F(x)| \to +\infty \tag{B.1.14}$$

when $x \to +\infty$.

If there exists some N such that A_k is nonnegative for $k \geq N$, then

$$\left| \sum_{k=N}^{\infty} A_k \rho^k x^k \right| \leq \sum_{k=N}^{\infty} A_k x^k$$

for any $x > 0$ and this contradicts (B.1.14) and proves the necessity of Condition A.

Sufficiency We use now the notations introduced in the beginning of this section. Then, from (B.1.5),

$$q_0^p(m) - q_k^p(m) = \sum_{j=1}^{p-1} \phi_j(m) a_{p-j}(q_0^j(m) - q_k^j(m)).$$

Letting

$$q_0(m) = r_0(1 + \varepsilon_0(m)), \qquad q_k(m) = r_k(1 + \varepsilon_k(m)),$$

since $\varepsilon_0(m)$ and $\varepsilon_k(m)$ tend to 0 when $m \to +\infty$, we have

$$p(\varepsilon_0(m) - \varepsilon_k(m)) = \sum_{j=1}^{p-1} \phi_j(m) a_{p-j}(1 - r_k^j) + o(\varepsilon_0(m) - \varepsilon_k(m))$$

when $m \to +\infty$ and with the definition of $\sigma(k)$

$$\varepsilon_0(m) - \varepsilon_k(m) \sim (1/p)\phi_{\sigma(k)}(m) a_{p-\sigma(k)}(1 - r_k^{\sigma(k)}).$$

From

$$\log\left|\frac{q_0(m)}{q_k(m)}\right| = \log\left|\frac{1 + \varepsilon_0(m)}{1 + \varepsilon_k(m)}\right| = \operatorname{Re}\left(\log\frac{1 + \varepsilon_0(m)}{1 + \varepsilon_k(m)}\right)$$
$$= \operatorname{Re}(\varepsilon_0(m) - \varepsilon_k(m)) + o(\varepsilon_0(m) - \varepsilon_k(m)),$$

we obtain that if the Condition A is satisfied,

$$|q_k(m)| < q_0(m)$$

$(k = 1, \ldots, p - 1)$ if m is great enough. Moreover

$$\log Q_0(m) - \log|Q_k(m)| = \sum_{j=M_0}^{m} \log q_0(m) - \log|q_k(m)|$$

and from (B.1.6) and the preceding relations

$$\log Q_0(m) - \log|Q_k(m)| \to +\infty$$

when $m \to +\infty$, so that

$$Q_k(m) = o(Q_0(m))$$

$(k = 1, \ldots, p - 1)$ when $m \to +\infty$. From Lemma B.1.1, we deduce that the coefficients $B_m = A_m \Gamma((m/p) + 1)$ satisfy

$$B_m = l_0 Q_0(m) + o(Q_0(m))$$

when $m \to +\infty$. If we prove that $l_0 \neq 0$, then all the B_m (and therefore all the A_m) are positive if m is great enough.

If the coefficients A_k of the power series

$$G(x) = \sum_{k=0}^{\infty} A_k x^k$$

satisfy (B.1.3), then G is solution of the differential equation

$$G'(x) - G(x)P'(x) = Q(x)$$

where Q is a polynomial with degree $p - 2$ and the solution of this equation is

$$y = e^{P(x)}\left(c - \int_x^{+\infty} e^{-P(t)}Q(t)\,dt\right)$$

which depends on p parameters, c, and the coefficients of Q, and we see easily that the coefficients l_k in (B.1.13) depend linearly on these parameters. If $c = 0$, the solution tends to 0 when $x \to +\infty$ and this implies an infinitely of change of signs in the sequence B_m, so that $l_0 = 0$. Therefore, $l_0 c^{-1}$ is a constant and if $c \neq 0$, $l_0 \neq 0$. But, for e^P, $c = 1$ and therefore $l_0 \neq 0$.

From (B.1.1) and (B.1.2), we obtain easily the relation

$$A_k = \sum \frac{a_1^{h_1}}{h_1!} \cdots \frac{a_p^{h_p}}{h_p!} \tag{B.1.15}$$

where the summation is taken on the $(h_1, \ldots, h_p) \in N^p$ satisfying

$$\sum_{j=1}^{p} jh_j = k.$$

From (B.1.15) and Theorem B.1.1, we obtain easily

Corollary B.1.1 If the set of real numbers $\{a_j : 1 \leq j \leq p\}$ satisfies Condition A, then nearly all the sums (B.1.15) are nonnegative.

We can refine the preceding result with

Theorem B.1.2 Let $\mathfrak{P}$ be a family of polynomials P

$$P(x) = \sum_{j=1}^{p} a_j x^j$$

satisfying the following conditions:

(a) P satisfies Condition A;
(b) the g.c.d. δ of $\{j : a_j > 0\}$ is equal to 1;
(c) for any j such that $a_j > 0$, there exist positive constants λ_j and μ_j such that

$$0 < \lambda_j \leq a_j \leq \mu_j < +\infty;$$

(d) for any j such that $a_j < 0$, there exist positive constants ν_j such that

$$|a_j| \le \nu_j .$$

For any $\varepsilon > 0$, there exists a number M depending only on ε, the λ_j, the μ_j, and the ν_j such that the coefficients A_k of the expansion

$$\exp(P(x)) = \sum_{k=0}^{\infty} A_k x^k$$

satisfy

$$0 < A_{m+1} < \varepsilon A_m$$

for any $m \ge M$ and any $P \in \mathfrak{P}$.

Proof We use again the notations introduced in the beginning of this section. Here

$$B_m = l_0 Q_0(m) + o(Q_0(m))$$

when $m \to +\infty$ and

$$\lim_{m\to\infty} \eta_0(m) = 1.$$

The number η which was used in the proof of Lemma B.1.1 can be any fixed number between 0 and $\frac{1}{2}$. The numbers M_0 [used for the validity of (B.1.7)], K_1, and M_1 [used for the validity of (B.1.8)], K_2, and M_2 [used for the validity of (B.1.9)] depend only on the coefficients of P and can be determined in function of the μ_j and the ν_j only. Moreover, if $\varepsilon = \log(1 + 1/2p)$, then $|c_k(m)/c_0(m)|$ tends monotonically to 0 up to the factor $1/2p$ if m is great enough and we can determine M_3 so that this is true for $m \ge M_3$ in functions of the μ_j and the ν_j only. Finally, we have introduced some M_4 such that $|Q_k(m)/Q_0(m)|$ tends monotonically to 0 for $m \ge M_4$ $(k = 1, \ldots, p-1)$. This number can be determined in functions of the λ_j, μ_j, and ν_j.

If we determine a number M greater than the M_j $(j = 0, 1, \ldots, 4)$ such that

$$|c_k(M)| < (1/2p)c_0(M) \tag{B.1.16}$$

$(k = 1, \ldots, p-1)$, then

$$|c_k(M)| < (1/p)c_0(M)$$

if $m \ge M$ and $B_m > 0$ for $m \ge M$.

The $c_k(m)$ are continuous functions of all the a_j. If we have (B.1.16) for some polynomial P, we have (B.1.16) for any polynomial deduced from P by a small variation of the nonnull coefficients. If the coefficients of P vary according to conditions (c) and (d) of the theorem, they are in a compact set Ω of a space with dimension p' (p' being the number of nonnull coefficients of P) and each point of this set belonging to an open set U for which we can determine a number M_U such that (B.1.16) holds for any point of U. Since Ω is compact, Ω can be covered by a finite number $U_1, \ldots, U_r$ of these domains. If

$$M = \sup_{1 \leq j \leq r} M_{U_j},$$

then we have $A_m > 0$ if $m \geq M$ for any $P \in \mathfrak{P}$. The second inequality can be proved by the same way, using the fact that

$$\lim_{m \to \infty} (B_{m+1})/B_m = 1.$$

Theorem B.1.3 Let P be the polynomial (B.1.1). If all the coefficients A_k of the expansion (B.1.2) are nonnegative, then P satisfies Conditions A and B.

Proof From Theorem B.1.1, we deduce that P satisfies Condition A. If P does not satisfy Condition B, let m be the smallest of the j such that a_j is negative and $j \notin (N)\mathfrak{B}_j$. Then $A_m = a_m$ and A_m is negative.

Theorem B.1.4 Let P_l be a polynomial

$$P_l(x) = \sum_{j=1}^{p} a_j(l)x^j$$

($l \geq 0$) satisfying the following conditions:

(a) P_l satisfies Conditions A and B for any l;
(b) the positive coefficients of P_l are constant;
(c) the absolute values of the negative coefficients are functions of l which are continuous, increasing, and tending to 0 when $l \to 0$ and one of them at least tends to $+\infty$ when $l \to +\infty$.

There exists some positive number l_0 such that all the coefficients $A_k(l)$ of the expansion

$$\exp(P_l(x)) = \sum_{k=0}^{\infty} A_k(l)x^k$$

are nonnegative if and only if $l \leq l_0$.

Proof If $0 \leq l \leq 1$, from Theorem B.1.2, we deduce the existence of some M such that $A_k(l) \geq 0$ if $k \geq M$. On the other hand, if $k < M$, since P_l satisfies Condition B, either $A_k \equiv 0$ or $\lim_{l \to 0} A_k(l) > 0$. Therefore, for l small enough, all the $A_k(l)$ are nonnegative.

Now let m be the smallest index such that the absolute value of the negative coefficient $a_m(l)$ tends to $+\infty$ when $l \to +\infty$. Since $A_m(l) - a_m(l)$ is bounded, A_m is negative for l great enough.

If, for some l, all the coefficients of the expansion of $\exp(P_l)$ are nonnegative, since $\exp(P_{l'})/\exp(P_l)$ is a power series with nonnegative coefficients for $l' < l$, all the coefficients of the expansion of $\exp(P_{l'})$ are also nonnegative. This implies the existence of a number l_0 such that all the coefficients of the expansion of $\exp(P_l)$ are nonnegative if $l < l_0$ and some of the coefficients of the expansion of $\exp(P_l)$ are nonnegative if $l > l_0$. Since the coefficients $A_k(l)$ are continuous functions of l, the coefficients of the expansion of $\exp(P_{l_0})$ are also nonnegative and the theorem is proved.

From Theorems B.1.3 and B.1.4, we deduce immediately

Corollary B.1.2 Let P be the polynomial (B.1.1). All the coefficients A_k of the expansion (B.1.2) are nonnegative if and only if the polynomial P satisfies Conditions A and B and the absolute values of those of the a_j which are negative are small enough.

From this result and (B.1.15), we have

Corollary B.1.3 If the set of real numbers $\{a_j : 1 \leq j \leq p\}$ satisfies Conditions A and B and if the absolute values of those of the a_j which are negative are small enough, then all the sums (B.1.15) are nonnegative.

B.2 CASE OF A FUNCTION OF ONE VARIABLE

Let P be the function

$$P(x) = \sum_{j=-p}^{q} a_j x^j \tag{B.2.1}$$

($p, q \in N$) and F be the expansion

$$F(x) = \exp(P(x)) = \sum_{k=-\infty}^{+\infty} A_k x^k \tag{B.2.2}$$

of e^P in Laurent series. We can suppose without loss of generality that $a_0 = 0$ and that the sets $\{j : j > 0,\ a_j \neq 0\}$ and $\{j : j < 0,\ a_j \neq 0\}$ are both nonempty. If we define P' and P'' by

$$P'(x) = \sum_{j=1}^{q} a_j x^j \quad \text{and} \quad P''(x) = \sum_{j=1}^{p} a_{-j} x^j,$$

then

$$P(x) = P'(x) + P''(x^{-1}).$$

We denote by δ' (resp. δ'') the g.c.d. of $\{j : j > 0, a_j \neq 0\}$ (resp. $\{-j : j < 0, a_j \neq 0\}$) and suppose that the g.c.d. of δ' and δ'' is equal to 1 (this does not restrict the generality).

If we define B_k and C_k by

$$\exp(P'(x)) = \sum_{k=0}^{\infty} B_k x^k, \qquad \exp(P''(x)) = \sum_{k=0}^{\infty} C_k x^k,$$

we have

$$A_k = \sum_{j=0}^{\infty} B_{j+k} C_j \tag{B.2.3a}$$

if $k > 0$ and

$$A_k = \sum_{j=0}^{\infty} C_{j-k} B_j \tag{B.2.3b}$$

if $k < 0$.

For any j, $-p \leq j \leq q$, $j \neq 0$, we introduce the following sets:

$$\mathfrak{A}_j = \{k : a_k \neq 0, k/j > 1\}, \qquad \mathfrak{A}'_j = \{k : a_k > 0, k/j > 1\},$$

$$\mathfrak{B}_j = \{k : a_k \neq 0, 0 < k/j < 1\}, \qquad \mathfrak{B}'_j = \{k : a_k > 0, 0 < k/j < 1\},$$

$$\mathfrak{C}_j = \{|k| : a_k \neq 0, k/j < 0\}, \qquad \mathfrak{C}'_j = \{|k| : a_k > 0, k/j < 0\},$$

and prove exactly as for Lemmas B.1.1 and B.1.2 the following results.

Lemma B.2.1 The following conditions are equivalent:

(A_1) for any $a_j < 0$, $j \in (Z)\mathfrak{A}_j$;
(A_2) for any $a_j < 0$, the g.c.d. of $\mathfrak{A}_j$ divides j;
(A_3) for any $a_j < 0$, $j \in (Z)\mathfrak{A}'_j$;
(A_4) for any $a_j < 0$, the g.c.d. of $\mathfrak{A}'_j$ divides j.

Lemma B.2.2 The following conditions are equivalent:

(C_1) for any $a_j < 0$, $j \in (N)$ $(\mathfrak{B}_j \cup \{\operatorname{sgn} j\rho_j\})$ where ρ_j is the g.c.d. of $\mathfrak{C}_j$;
(C_2) for any $a_j < 0$, $j \in (N)$ $(\mathfrak{B}'_j \cup \{\operatorname{sgn} j\rho_j\})$.

It must be noted that ρ_j is equal to δ'' (resp. δ') if j is positive (resp. negative).

Lemma B.2.3 The following conditions are equivalent:

(D_1) for any $a_j < 0$, $j \in (N)$ $(\mathfrak{B}_j \cup \{\operatorname{sgn} j\rho_j\})\backslash(N)\mathfrak{B}_j$;
(D_2) for any $a_j < 0$, $j \in (N)$ $(\mathfrak{B}'_j \cup \{\operatorname{sgn} j\rho_j\})\backslash(N)\mathfrak{B}'_j$.

We will call Condition A (resp. Condition C, Condition D) one of the conditions of Lemma B.2.1 (resp. B.2.2, B.2.3). We have also the easily proved

Lemma B.2.4 If P satisfies Condition A, then for any j

$$\rho_j = \rho'_j$$

where ρ'_j is the g.c.d. of $\mathfrak{C}'_j$.

For any $m \geq 0$, we define $\rho(m)$ and $\sigma(m)$ by

$$\rho(m) = \inf\{j : j + m \in (N)\{\delta'\},\ C_j \neq 0\},$$

$$\sigma(m) = \inf\{j : j + m \in (N)\{\delta''\},\ B_j \neq 0\}.$$

From Bezout's relation, we deduce easily the existence of $\rho(m)$ and $\sigma(m)$. We have evidently the relations

$$\rho(0) = \sigma(0) = 0, \qquad \rho(m + \delta') = \rho(m), \qquad \sigma(m + \delta'') = \sigma(m).$$

Lemma B.2.5 We suppose that the function P defined by (B.2.1) satisfies Condition A and that the sets $\{j : j > 0,\ a_j \neq 0\}$ and $\{j : j < 0,\ a_j \neq 0\}$ are both nonempty. All the coefficients A_k of expansion (B.2.2) are nonnegative except a finite number if and only if

$$C_{\rho(m)} > 0, \qquad m = 1, 2, \ldots, \delta' - 1,$$

$$B_{\sigma(m)} > 0, \qquad m = 1, 2, \ldots, \delta'' - 1.$$

Proof If P satisfies Condition A, it follows from Theorem B.1.2 that for any $\varepsilon > 0$, there exists some $M(\varepsilon)$ such that

$$0 < B_{(m+1)\delta'} < \varepsilon B_{m\delta'}$$

if $m \geq M(\varepsilon)$. If $k > 0$, then from (B.2.3a) and the definition of $\rho(m)$, we have

$$A_k = B_{k+\rho(k)} C_{\rho(k)} + \sum_{j=1}^{\infty} B_{k+\rho(k)+j\delta'\delta''} C_{\rho(k)+j\delta'\delta''}$$

since B_l (resp. C_m) is null if δ' (resp. δ'') does not divide l (resp. m). From this, we deduce easily that if $\varepsilon < 1$ and k is great enough $[k \geq M(\varepsilon)\delta']$, then

$$|A_k - B_{k+\rho(k)} C_{\rho(k)}| < B_{k+\rho(k)} |C_{\rho(k)}| (\eta\varepsilon/(1 - \varepsilon))$$

where

$$\eta = \sup|C_m| / \inf|C_{\rho(m)}|$$

and A_k has the sign of $C_{\rho(k)}$ if $\varepsilon < 1/(1 + \eta)$ and $k \geq M(\varepsilon)\delta'$.

In the same way, for any $\varepsilon' > 0$, there exists some $M'(\varepsilon')$ such that

$$0 < C_{(m+1)\delta''} < \varepsilon' C_{m\delta''}$$

if $m \geq M'(\varepsilon')$. If $\eta' = \sup|B_m| / \inf|B_{\sigma(m)}|$ and $\varepsilon' < 1/(1 + \eta')$, we deduce from (B.2.3b) that A_k has the sign of $B_{\sigma(k)}$ if $-k \geq M'(\varepsilon')\delta''$. From this, we deduce easily the lemma.

Theorem B.2.1 Let P be the function defined by (B.2.1). If $\{j : j > 0, a_j \neq 0\}$ and $\{j : j < 0, a_j \neq 0\}$ are both nonempty and if all the coefficients A_k of the expansion (B.2.2) are nonnegative except a finite number, then P satisfies Conditions A and C.

Proof For Condition A, the proof is exactly the same as the one for Theorem B.1.1. If there exists some $m > 0$ (resp. $m < 0$) such that $a_m < 0$ and $m \notin (N)\ (\mathfrak{B}_m \cup \{\delta''\})$ [resp. $m \notin (N)\ (\mathfrak{B}_m \cup \{-\delta'\}]$, then m is one of the $\sigma(k)$ [resp. the opposite of one of the $\rho(k)$] and $B_m = a_m$ (resp. $C_{-m} = a_m$). Therefore, if P satisfies Condition A and does not satisfy Condition C, it follows from Lemma B.2.5 that the set $\{k : A_k < 0\}$ is infinite.

Theorem B.2.2 Let P_l be the function defined by

$$P_l(x) = \sum_{j=-p}^{q} a_j(l)x^j \tag{B.2.4}$$

$(l \geq 0)$ and

$$\exp(P_l(x)) = \sum_{k=-\infty}^{+\infty} A_k(l)x^k. \tag{B.2.5}$$

If the following conditions are satisfied:

(a) P_l satisfies Conditions A and C for $l \geq 0$ and does not satisfy Condition D for $l > 0$;
(b) the sets $\{j : j > 0, a_j \neq 0\}$ and $\{j : j < 0, a_j \neq 0\}$ are both nonempty;
(c) the positive coefficients of P_l are constant;
(d) the absolute values of the negative coefficients of P_l are continuous increasing functions of l;
(e) the absolute values of negative coefficients of P_l which do not satisfy Condition D tend to zero when $l \to 0$ and one of them at least tends to $+\infty$ when $l \to +\infty$,

there exist some positive constants l_0 and l'_0 such that the set $\{k : A_k < 0\}$ is finite (even the number of its elements is bounded by a constant independent of l) if $l < l_0$ and is infinite if $l > l'_0$.

Proof Let m be the smallest positive (resp. negative) integer such that m does not satisfy Condition D and $|a_m|$ tends to $+\infty$ when $l \to +\infty$. Then m is one of the $\sigma(k)$ [resp. the opposite of one of the $\rho(k)$] and B_m (resp. C_{-m}) is negative if $|a_m|$ is great enough. Therefore, if l is great enough $(l > l'_0)$, it follows from Lemma B.2.5 that $\{k : A_k > 0\}$ is infinite.

Now, if P_l satisfies Condition C, for any k, $B_{\sigma(k)}$ and $C_{\rho(k)}$ tend to a positive limit when $l \to 0$ and, since there are only a finite number of these functions to consider, there exists some l_0 such that $B_{\sigma(k)}$ and $C_{\rho(k)}$ are positive if $l < l_0$.

Since P_l satisfies Condition A, we obtain from Lemma B.2.5 that $\{k : A_k < 0\}$ is finite if $l < l_0$. If we look at the proof of Lemma B.2.5, we find that the number of elements of the set $\{k : A_k < 0\}$ can be bounded independently of l if l is small enough.

For $l = 0$, we have the

Corollary B.2.1 If the function P defined by (B.2.1) satisfies Conditions A and D, then all the coefficients A_k of the expansion (B.2.2) except a finite number are nonnegative.

Theorem B.2.3 Let P_l be the function defined by (B.2.4) ($l \geq 0$). If the following conditions are satisfied:

(a) P_l satisfies Conditions A and C for any l;
(b) the positive coefficients of P_l are constant;
(c) the absolute values of the negative coefficients of P_l are continuous increasing functions of l tending to zero when $l \to 0$ and one of them at least tends to $+\infty$ when $l \to +\infty$,

there exists some positive constant l_0 such that all the coefficients A_k of expansion (B.2.5) are nonnegative if and only if $l \leq l_0$.

The proof is very similar to the one of Theorem B.1.4 and can be omitted. From Theorems B.2.1 and B.2.3, we obtain immediately

Corollary B.2.2 Let P be function (B.2.1). All the coefficients A_k of expansion (B.2.2) are nonnegative if and only if the function P satisfies Conditions A and C and if the absolute values of those of the a_j which are negative are small enough.

From (B.2.1) and (B.2.2), we obtain

$$A_k = \sum \frac{a_{-p}^{h_{-p}}}{h_{-p}!} \cdots \frac{a_{-1}^{h_{-1}}}{h_{-1}!} \frac{a_1^{h_1}}{h_1!} \cdots \frac{a_q^{h_q}}{h_q!} \tag{B.2.6}$$

where the summation is taken on the $(h_{-p}, \ldots, h_{-1}, h_1, \ldots, h_q) \in N^{p+q}$ satisfying

$$\sum_{j=-p}^{q} jh_j = k.$$

From this and Corollary B.2.2, we have

Corollary B.2.3 If the set of real numbers $\{a_j : -p \leq j < q, j \neq 0\}$ satisfies Conditions A and C and if the absolute values of those of the a_j which are negative are small enough, then all the sums (B.2.6) are nonnegative.

B.3 CASE OF FUNCTIONS OF SEVERAL VARIABLES

Theorem B.3.1 Let $x \in R^n$, $p \in N^n$, and P be the polynomial

$$P(x) = \sum_{0 \le k \le p} a_k x^k \tag{B.3.1}$$

($a_0 = 0$). All coefficients A_k of the expansion

$$\exp(P(x)) = \sum_{k \in N^n} A_k x^k \tag{B.3.2}$$

except a finite number are nonnegative if and only if the polynomial Q of the real variable y defined by

$$Q(y) = P(y^{\rho_1}, \ldots, y^{\rho_n})$$

satisfies Condition A for any $(\rho_1, \ldots, \rho_n) \in N^n$.

Proof The necessity is clear. We prove now the sufficiency. If for some m, a_m is negative, for any $\rho = (\rho_1, \ldots, \rho_n) \in N^n$, there exist $k_1, \ldots, k_l \in N^n$ and some integers $\alpha_1(\rho), \ldots, \alpha_l(\rho)$ such that

$$a_{k_j} \neq 0$$

($j = 1, \ldots, l$) and

$$(m, \rho) = \sum_{j=1}^{l} \alpha_j(\rho)(k_j, \rho).$$

Since this is true for any ρ, we deduce easily that we can choose the α_j independently of ρ. We have then

$$m = \sum_{j=1}^{l} \alpha_j k_j .$$

From (B.3.1) and (B.3.2), we have

$$A_k = \sum \frac{a_{k_1}^{h_1}}{h_1!} \cdots \frac{a_{k_r}^{h_r}}{h_r!}$$

where the summation is taken on the $(h_1, \ldots, h_r) \in N^p$ such that

$$\sum_{j=1}^{r} h_j k_j = k$$

(r is the number of nonnull coefficients in P). But, from Corollary B.1.1, we deduce easily that nearly all these sums are nonnegative.

With the same proof, we deduce from Corollaries B.1.3 and B.2.3 the following results.

Theorem B.3.2 Let P be polynomial (B.3.1). All the coefficients A of expansion (B.3.2) are nonnegative if and only if the two following conditions are satisfied:

(a) the polynomial Q defined by

$$Q(y) = P(y^{\rho_1}, \ldots, y^{\rho_n})$$

satisfies Conditions A and B for any $(\rho_1, \ldots, \rho_n) \in N^n$;

(b) the absolute values of those of the a_j which are negative are small enough.

Theorem B.3.3 Let $x \in R^n$, $p \in N^n$, $q \in N^n$, and P be the function

$$P(x) = \sum_{-p \leq j \leq q} a_j x^j.$$

Then all the coefficients A_k of the expansion

$$\exp(P(x)) = \sum_{k \in Z^n} A_k x^k$$

are nonnegative if and only if the two following conditions are satisfied:

(a) the function Q defined by

$$Q(y) = P(y^{\rho_1}, \ldots, y^{\rho_n})$$

satisfies Conditions A and C for any $(\rho_1, \ldots, \rho_n) \in Z^n$;

(b) the absolute values of those of the a_j which are negative are small enough.

References

Aczél, J. (1966). "Lectures on Functional Equations and Their Applications." Academic Press, New York.

Bary, N. K. (1964). "A Treatise on Trigonometric Series." Pergamon, Oxford.

Bochner, S. (1933). Monotone Funktionen, Stieltjessche Integrale und harmonische Analyse. *Math. Ann.* **108**, 378–410.

Bochner, S. (1934). A theorem on Fourier–Stieltjes integrals. *Bull. Amer. Math. Soc.* **40**, 271–276.

Bochner, S., and Chandrasekharan, K. (1949). "Fourier Transforms" (Ann. Math. Studies No. 19). Princeton Univ. Press, Princeton, New Jersey.

Cauchy, A. (1853a). Mémoire sur les coefficients limitateurs ou restricteurs. *C. R. Acad. Sci. Paris* **37**, 150–152 ["Oeuvres Complètes," 1st Series, Vol. 12, pp. 79–94. Gauthier-Villars, Paris].

Cauchy, A. (1853b). Sur les résultats moyens d'observations de même nature et sur les résultats les plus probables. *C. R. Acad. Sci. Paris* **37**, 198–206 ["Oeuvres Complètes," 1st Series, Vol. 12, pp. 94–104. Gauthier-Villars, Paris].

Cauchy, A. (1853c). Sur la probabilité des erreurs qui affectent des résultats moyens d'observations de même nature. *C. R. Acad. Sci. Paris* **37**, 264–272 ["Oeuvres Complètes," 1st Series, Vol. 12, pp. 104–114. Gauthier-Villars, Paris].

Choquet, G. (1969). "Lectures on Analysis." Benjamin, New York.

Chung, K. L. (1968). "A Course in Probability Theory." Harcourt, New York.

Čistjakov, G. P. (1971). On the belonging to the class I_0 of laws with nonanalytic characteristic functions. *Dokl. Akad. Nauk SSSR* **201**, 280–283 [Engl. transl.: *Soviet Math. Dokl.* **12**, 1654–1658].

Cramér, H. (1936). Uber eine Eigenschaft der normalen Verteilungsfunktion. *Math. Z.* **41**, 405–414.

Cramér, H. (1937). "Random Variables and Probability Distributions." Cambridge Univ. Press, London and New York.

Cramér, H. (1939). On the representation of a function by certain Fourier integrals. *Trans. Amer. Math. Soc.* **46**, 191–201.

Cramér, H. (1949). On the factorization of certain probability distributions. *Ark. Mat.* **1**, 61–65.

Cramér, H., and Wold, H. (1936). Some theorems on distribution functions. *J. London Math. Soc.* **11**, 290–294.

Cuppens, R. (1963a). Sur la décomposition d'une fonction 2q fois dérivable à l'origine en produit infini de fonctions caractéristiques. *C. R. Acad. Sci. Paris* **256**, 3806–3808.

Cuppens, R. (1963b). Sur un théorème de Mamay. *C. R. Acad. Sci. Paris* **257**, 586–588.

Cuppens, R. (1967). Décomposition des fonctions caractéristiques des vecteurs aléatoires. *Publ. Inst. Statist. Univ. Paris* **16**, 61–153.

Cuppens, R. (1968). Sur un théorème de Paul Lévy. *Publ. Inst. Statist. Univ. Paris* **17**, No. 3, 1–6.

Cuppens, R. (1969a). On the decomposition of infinitely divisible characteristic functions of several variables. *Z. Wahrscheinlichkeitstheorie und Verw. Gebiete* **12**, 59–72.

Cuppens, R. (1969b). On the decomposition of infinitely divisible probability laws without normal factor. *Pacific J. Math.* **28**, 61–76.

Cuppens, R. (1969c). On finite products of Poisson-type characteristic functions of several variables. *Ann. Math. Statist.* **40**, 434–444.

Cuppens, R. (1969d). On the decomposition of infinitely divisible characteristic functions with continuous Poisson spectrum. *Proc. Amer. Math. Soc.* **21**, 145–152.

Cuppens, R. (1969e). On the decomposition of infinitely divisible characteristic functions with continuous Poisson spectrum. II. *Pacific J. Math.* **29**, 521–525.

Cuppens, R. (1969f). Décomposition des fonctions caractéristiques indéfiniment divisibles de plusieurs variables à spectre de Poisson continu. *Ann. Inst. H. Poincaré Sect. B* **5**, 123–133.

Cuppens, R. (1969g). Quelques nouveaux résultats en arithmétique des lois de probabilité. *In C. R. Colloq. C.N.R.S.* "Les probabilités sur les structures algébriques," pp. 97–112. C.N.R.S., Paris.

Cuppens, R. (1970). Application de la théorie des fonctions caractéristiques de plusieurs variables à l'étude de certains problèmes de décomposition des fonctions caractéristiques d'une variable. *Z. Wahrscheinlichkeitstheorie und Verw. Gebiete* **15**, 144–156.

Cuppens, R. (1971). Sur un problème de R. Shimizu, *J. Multivariate Anal.* **1**, 276–287.

Cuppens, R. (1972a). Isomorphismes et arithmétique des semigroupes de lois de probabilité. *Z. Wahrscheinlichkeitstheorie und Verw. Gebiete* **21**, 147–153.

Cuppens, R. (1972b). Independent sets and factorization of probability laws. *J. Multivariate Anal.* **2**, 239–248.

Cuppens, R. (1973). An isomorphism method for the study of I_0^n. *In* "Multivariate Analysis" (P. R. Krishnaiah, ed.). Vol. III. pp. 183–197. Academic Press, New York.

Davidson, R. (1968). Arithmetic and other properties of certain Delphic semi-groups. *Z. Wahrscheinlichkeitstheorie und Verw. Gebiete* **10**, 120–145, 146–172.

Davidson, R. (1969). More delphic theory and practice. *Z. Wahrscheinlichkeitstheorie und Verw. Gebiete* **13**, 191–203.

De Finetti, B. (1929). Sulla funzione a incremento aleatorio. *Atti Accad. Naz. Lincei Rend. Cl. Sci. Fis. Mat. Natur.* **10**, 163–168, 325–329, 548–553.

Dieudonné, J. (1969). "Foundations of Modern Analysis," 2nd ed. Academic Press, New York.

Dugué, D. (1951a). Analyticité et convexité des fonctions caractéristiques. *Ann. Inst. H. Poincaré* **12**, 45–56.

Dugué, D. (1951b). Sur certains exemples de décomposition en arithmétique des lois de probabilité. *Ann. Inst. H. Poincaré* **12**, 159–169.

Dugué, D. (1956). "Arithmétique des lois de probabilité" [*Mémor. Sci. Math.* **No. 137**]. Gauthier-Villars, Paris.

Dugué, D. (1957). Sur le théorème de Lévy-Cramér. *Publ. Inst. Statist. Univ. Paris* **6**, 213–225.

Dugué, D., and Girault, M. (1955). Fonctions convexes de Pólya. *Publ. Inst. Statist. Univ. Paris* **4**, 3–10.
Erdös, P., and Stone, A. H. (1970). On the sum of two Borel sets. *Proc. Amer. Math. Soc.* **25**, 304–306.
Esseen, C. G. (1965). On infinitely divisible one-sided distributions. *Math. Scand.* **17**, 65–76.
Favard, J. (1933). "Leçons sur les fonctions presque-périodiques." Gauthier-Villars, Paris.
Feller, W. (1966). "An Introduction to Probability Theory and Its Applications," Vol. II. Wiley, New York.
Fuks, B. A. (1963). "Introduction to the Theory of Analytic Functions of Several Complex Variables" [Transl. of Math. Monogr., No. 8]. Amer. Math. Soc., Providence, Rhode Island.
Gelfand, I., Raikov, D., and Shilov, G. (1964). "Commutative Normed Rings." Chelsea, Bronx, New York.
Gilbert, W. M. (1955). Projections of probability distributions. *Acta Math. Acad. Sci. Hungar.* **6**, 195–198.
Gnedenko, B. V. (1939). On the theory of limit theorems for sums of independent random variables. *Izv. Akad. Nauk SSSR Ser. Mat.* **3**, 181–232, 643–647.
Gndenko, B. V. (1944). Elements of the theory of distribution functions of random vectors. *Uspehi Mat. Nauk* **10**, 230–244.
Goldberg, A. A., and Ostrovskiĭ, I. V. (1967). Application of a theorem of W. K. Hayman to a question in the theory of decompositions of probability laws. *Ukrain. Mat. Ž.* **19**, No. 3, 104–106 [Engl. Transl.: *Select. Transl. Math. Statist. Probability* **9**, 147–151 (1971)].
González Domínguez, A. (1940). The representation of functions by Fourier integrals. *Duke Math. J.* **6**, 246–255.
Halmos, P. (1950). "Measure Theory." Van Nostrand-Reinhold, Princeton, New Jersey.
Hardy, G. H., Littlewood, J. E., and Pólya, G. (1952). "Inequalities," 2nd ed. Cambridge Univ. Press, London and New York.
Haviland, E. K. (1934a). On the theory of absolutely additive distribution functions. *Amer. J. Math.* **56**, 625–658.
Haviland, E. K. (1934b). On distribution functions and their Laplace-Fourier transforms. *Proc. Nat. Acad. Sci. U.S.A.* **20**, 50–57.
Haviland, E. K. (1935a). On the inversion formula for Fourier–Stieltjes transforms in more than one dimension. *Amer. J. Math.* **57**, 94–100, 382–388.
Haviland, E. K. (1935b). A note on a property of Fourier–Stieltjes transforms in more than one dimension. *Amer. J. Math.* **57**, 569–572.
Haviland, E. K. (1937). On the expression of a given function as the Fourier–Stieltjes transform of a distribution function whose spectrum is confined to a given set. *J. London Math. Soc.* **12**, 253–257.
Heppes, A. (1956). On the determination of probability distributions of more dimensions by their projections. *Acta Math. Acad. Sci. Hungar.* **7**, 403–410.
Hinčin, A. Ja. (1937a). A new deduction of P. Lévy's formula (in Russian). *Bull. Moscow Gos. Univ. Sect. A1* **1**, 1–5.
Hinčin, A. Ja. (1937b). The arithmetic of distribution laws (in Russian). *Bull. Moscow Gos. Univ. Sect. A1* **1**, 6–17.
Johansen, S. (1966). An application of extreme point methods to the representation of infinitely divisible distributions. *Z. Wahrscheinlichkeitstheorie und Verw. Gebiete* **5**, 304–316.
Kagan, A. M., Linnik, Ju. V., and Rao, S. R. (1972). "Characterization Problems of Mathematical Statistics" (in Russian). Izdat. Nauka, Moscow.
Kahane, J. P., and Salem, R. (1963). "Ensembles parfaits et séries trigonométriques." Hermann, Paris.
Kawata, T. (1950). Representation of a function by the Fourier–Stieltjes integral. *Ann. Inst. Statist. Math.* **1**, 131–139.

Kendall, D. G. (1963). Extreme-point methods in stochastic analysis. *Z. Wahrscheinlichkeitstheorie und Verw. Gebiete* **1**, 295–300.

Kendall, D. G. (1968). Delphic semi-groups, infinitely divisible regenerative phenomena and the arithmetic of p-functions. *Z. Wahrscheinlichkeitstheorie und Verw. Gebiete* **9**, 163–195.

Kolmogorov, A. N. (1932). Sulla forma generale di un processo stocastico omogeneo. *Atti Accad. Naz. Lincei Rend. Cl. Sci. Fis. Mat. Natur.* **15**, 805–808, 866–869.

Kotlarski, I. (1971). On a characterization of probability distributions by the joint distribution of some of their linear forms. *Sankhyā Ser. A* **33**, 73–80.

Kuratowski, C. (1952). "Topologie," 3rd ed., Vol. I. [Monogr. Mat. Tom **20**]. PańStowe Wydawnictivo Naukowe, Warsaw.

Laha, R. G. (1958). On a factorization theorem in the theory of analytic characteristic functions. *Ann. Math. Statist.* **29**, 922–926.

Laha, R. G. (1969). On an analytical decomposition of the Poisson law. *Trans. Amer. Math. Soc.* **140**, 137–148.

Laha, R. G., and Lukacs, E. (1962). On a factorization of characteristic functions which have a finite number of derivatives at the origin. *Publ. Inst. Statist. Univ. Paris* **11**, 221–224.

Laplace, P. S. (1795). Théorie analytique des probabilités. ["Oeuvres Complétes," Vol. 7, Gauthier-Villars, Paris.]

Lepetit, C. (1973). Décomposition de distributions à support compact. Application à la factorisation de certaines lois de probabilité. *Z. Wahrscheinlichkeitstheorie und Verw. Gebiete* **26**, 51–65.

Letta, G. (1963). Eine Bemerkung zur Kennzeichnung der charakteristischen Funktionen. *Z. Wahrscheinlichkeitstheorie und Verw. Gebiete* **2**, 69–74.

Lévy, P. (1925). "Calcul des probabilités." Gauthier-Villars, Paris.

Lévy, P. (1934). Sur les intégrales dont les éléments sont des variables aléatoires indépendantes. *Ann. Scuola Norm. Sup. Pisa* **3**, 337–366.

Lévy, P. (1937a). Sur les exponentielles de polynômes et sur l'arithmétique des produits de lois de Poisson. *Ann. Sci. École Norm. Sup.* **54**, 231–292.

Lévy, P. (1937b). "Théorie de l'addition des variables aléatoires." Gauthier-Villars, Paris.

Lévy, P. (1938a). L'arithmétique des lois de probabilité. *J. Math. Pures Appl.* **17**, 17–39.

Lévy, P. (1938b). L'arithmétique des lois de probabilité et les produits finis de lois de Poisson. *Actualités Sci. Indust.* **No. 736**, 25–59. Hermann, Paris.

Lévy, P. (1948). The arithmetical character of the Wishart distribution. *Proc. Cambridge Philos. Soc.* **44**, 295–297.

Lévy, P. (1968). Observations sur la note précédente. *C. R. Acad. Sci. Paris Sér. A* **266**, 728–729.

Lewis, T. (1967). The factorization of the rectangular distribution. *J. Appl. Probability* **4**, 529–542.

Linnik, Ju. V. (1955). A problem on characteristic functions of probability distributions. *Uspehi Mat. Nauk* **10**, 137–138.

Linnik, Ju. V. (1956). A remark on Cramér's theorem on the decomposition of the normal law. *Teor. Verojatnost. i Primenen.* **1**, 479–480 [Engl. Transl.: *Theor. Probability Appl.* **1**, 435–436].

Linnik, Ju. V. (1957). On the decomposition of the convolution of Gauss and Poissonian laws. *Teor. Verojatnost. i Primenen.* **2**, 34–59 [Engl. Transl.: *Theor. Probability Appl.* **2**, 31–57].

Linnik, Ju. V. (1958). General theorems on the factorization of infinitely divisible laws. I. *Teor. Verojatnost. i Primenen.* **3**, 3–40 [Engl. Transl.: *Theor. Probability Appl.* **3**, 1–37].

Linnik, Ju. V. (1959a). General theorems on the factorization of infinitely divisible laws. II. *Teor. Verojatnost. i Primenen.* **4**, 55–85 [Engl. Transl.: *Theor. Probability Appl.* **4**, 53–82].

Linnik, Ju. V. (1959b). General theorems on the factorization of infinitely divisible laws. III. *Teor. Verojatnost. i Primenen.* **4**, 150–171 [Engl. Transl.: *Theor. Probability Appl.* **4**, 142–163].

Linnik, Ju. V. (1959c). On "α-factorizations" of infinitely divisible probability laws. *Vestnik Leningrad. Univ.* **14**, No. 1, 14–23 [Engl. Transl.: *Select. Transl. Math. Statist. Probability* **2**, 159–169 (1962)].

Linnik, Ju. V. (1960). "Decomposition of Probability Distribution," (in Russian) Izdat. Leningrad. Univ., Leningrad [French Transl.: Gauthier-Villars, Paris, 1962; Engl. Transl.: Dover, New York, and Oliver & Boyd, Edinburgh, 1964].

Linnik, Ju. V., and Ostrovskiĭ, I. V. (1972). "Decomposition of Random Variables and Vectors" (in Russian). Izdat. Nauka, Moscow.

Livsič, L. Z. (1970). A sufficient condition for a two-variable infinitely divisible law to have only infinitely divisible components. *Teor. Funkciĭ Funkcional. Anal. i Priložen.* **12**, 36–59.

Livsič, L. Z. (1972). On the belonging to the class I_{on} of infinitely divisible laws. *Dopovidi Akad. Nauk Ukrain. RSR Ser. A* **No. 11**, 992–993.

Livsič, L. Z., and Ostrovskiĭ, I. V. (1971). On the multidimensional infinitely divisible laws that have only infinitely divisible components. *Dokl. Akad. Nauk SSSR* **198**, 1273 [Engl. Transl.: *Soviet Math. Dokl.* **12**, 978–979].

Lukacs, E. (1952). An essential property of the Fourier transform of distribution functions. *Proc. Amer. Math. Soc.* **3**, 508–510.

Lukacs, E. (1957). Les fonctions caractéristiques analytiques. *Ann. Inst. H. Poincaré* **15**, 217–251.

Lukacs, E. (1964). A linear mapping of the space of distribution functions onto a set of bounded continuous functions. *Z. Wahrscheinlichkeitstheorie und Verw. Gebiete* **3**, 1–6.

Lukacs, E. (1969). Some recent developments in the theory of multivariate characteristic functions. *In* "Multivariate Analysis" (P. R. Krishnaiah, ed.), Vol. II, pp. 303–319. Academic Press, New York.

Lukacs, E. (1970). "Characteristic Functions," 2nd ed. Griffin, London.

Lukacs, E., and Laha, R. G. (1964). "Applications of Characteristic Functions" (Griffin's statistical monographs and courses, No. 14). Hafner, New York.

Lusin, N. (1930). "Leçons sur les ensembles analytiques et leurs applications." Gauthier-Villars. Paris.

Mamaĭ, L. V. (1960). On the theory of characteristic functions. *Vestnik Leningrad. Univ.* **15**, No. 1, 85–99 [Engl. Transl.: *Select. Transl. Math. Statist. Probability* **4**, 153–170 (1963)].

Marcinkiewicz, J. (1938). Sur les fonctions indépendantes, III. *Fund. Math.* **31**, 86–102.

Marcus, M., and Minc, H. (1965). "Introduction to Linear Algebra." Macmillan, New York.

Morucci, B. (1966). Existence de la dérivée première d'une fonction caractéristique. Généralisation aux dérivées d'ordre impair. *C. R. Acad. Sci. Paris Sér. A* **263**, 320–323.

Ostrovskiĭ, I. V. (1964). Decomposition of infinitely divisible lattice laws. *Vestnik Leningrad. Univ.* **19**, No. 4, 51–60 [Engl. Transl.: *Select. Transl. Math. Statist. Probability* **9**, 127–139 (1971)].

Ostrovskiĭ, I. V. (1965a). Some theorems on decomposition of probability laws. *Trudy Mat. Inst. Steklov* **79**, 198–235 [Engl. Transl.: *Proc. Steklov Inst. Math.* **79**, 221–259].

Ostrovskiĭ, I. V. (1965b). On factorizing the composition of a Gauss and a Poisson distribution. *Uspehi Mat. Nauk* **20**, No. 4, 166–171.

Ostrovskiĭ, I. V. (1965c). A multidimensional analogue of Ju. V. Linnik's theorem on decomposition of a Gauss and a Poisson law. *Teor. Verojatnost. i Primenen.* **10**, 742–745 [Engl. Transl.: *Theor. Probability Appl.* **10**, 673–677].

Ostrovskiĭ, I. V. (1966a). Certain properties of holomorphic characteristic functions of multidimensional probability laws. *Teor. Funkciĭ Funkcional. Anal. i Priložen.* **2**, 169–177.

Ostrovskiĭ, I. V. (1966b). Decomposition of multidimensional infinitely divisible laws without Gaussian component. *Vestnik Har'kov. Gos. Univ.* **32**, No. 14, 51–72.

Ostrovskiĭ, I. V. (1970). On some classes of infinitely divisible laws. *Izv. Akad. Nauk SSSR Ser. Mat.* **34**, 923–944.

Ostrovskiĭ, I. V. (1972). On the decomposition theory of multi-dimensional infinitely divisible laws. *Dopovidi Akad. Nauk Ukrain. RSR Ser. A* **No. 11**, 997–1000.

Paley, R. E. A. C., and Wiener, N. (1934). "Fourier Transforms in the Complex Domain" (*Amer. Math. Soc. Colloq. Publ.* **No. 19**). Amer. Math. Soc., Providence, Rhode Island.

Parthasarathy, K. R. (1967). "Probability Measures on Metric Spaces." Academic Press, New York.

Parthasarathy, K. R., Ranga Rao, R., and Varadhan, S. R. (1962). On the category of indecomposable distributions on topological groups. *Trans. Amer. Math. Soc.* **102**, 200–217.

Pitman, E. J. G. (1956). On the derivation of a characteristic function at the origin. *Ann. Math. Statist.* **27**, 1156–1160.

Plancherel, M., and Pólya, G. (1937). Fonctions entières et intégrales de Fourier multiples. *Comment. Math. Helv.* **9**, 224–248.

Poincaré, H. (1912). "Calcul des probabilités." Carré-Naud, Paris.

Pólya, G. (1949). Remarks on characteristic functions. *Proc. Berkeley Symp. Math. Statist. Probability, Berkeley, 1949*, pp. 115–123.

Raĭkov, D. A. (1937). Decomposition of Poisson laws. *Dokl. Akad. Nauk SSSR* **14**, 9–12.

Raĭkov, D. A. (1938a). Decomposition of Gauss and Poisson laws. *Izv. Akad. Nauk SSSR Ser. Mat.* **2**, 91–124.

Raĭkov, D. A. (1938b). A theorem in the theory of analytic characteristic functions. *Izv. Nauchno Issled. Mat. Meh. Tomsk. Gos. Univ.* **2**, No. 2, 8–11.

Ramachandran, B. (1964). Application of a theorem of Mamay's to a "denumerable α-decomposition" of the Poisson law. *Publ. Inst. Statist. Univ. Paris* **13**, 13–19.

Ramachandran, B. (1965). An extension of a theorem of Mamay with application. *Sankhyā Ser. A* **27**, 303–310.

Ramachandran, B. (1966). On one-sided distribution functions. *Sankhyā Ser. A* **28**, 315–318.

Ramachandran, B. (1967). "Advanced Theory of Characteristic Functions." Statist. Publ. Soc., Calcutta.

Rényi, A. (1952). On projections of probability distributions. *Acta Math. Acad. Sci. Hungar.* **3**, 131–142.

Riesz, F. (1933). Uber Sätze von Stone und Bochner. *Acta Litt. Sci. Szeged* **6**, 184–198.

Rousseau, B. (1973). Arithmétique des lois de probabilité définies sur un espace de Hilbert séparable. *C. R. Acad. Sci. Paris Sér. A* **276**, 141–143.

Rousseau, B. (1974). Arithmétique des lois de probabilité définies sur un espace de Hilbert séparable. *J. Multivariate Anal.* **4**, 1–21.

Rvačeva, E. L. (1954). On domains of attraction of multidimensional distributions. *L'vov. Gos. Univ. Uč. Zap. Ser. Meh. Mat.* **29**, No. 6, 5–44 [Engl. Transl.: *Select. Transl. Math. Statist. Probability* **2**, 183–205 (1962)].

Schoenberg, I. J. (1934). A remark on the preceding note by Bochner. *Bull. Amer. Math. Soc.* **40**, 277–278.

Schoenberg, I. J. (1938). Metric spaces and positive-definite functions. *Trans. Amer. Math. Soc.* **44**, 522–536.

Schwartz, L. (1959). "Etude des sommes d'exponentielles" [*Actualités Sci. Indust.* **No. 959**]. Hermann, Paris.

Shimizu, R. (1964). On the decomposition of infinitely divisible characteristic functions with a continuous Poisson spectrum. *Ann. Inst. Statist. Math.* **16**, 387–407.

Teicher, H. (1954). On the multivariate Poisson distribution. *Skand. Aktuarietidskr.* **37**, 1–9.

Titchmarsh, E. C. (1939). "The Theory of Functions." 2nd ed. Oxford Univ. Press, London and New York.

Tortrat, A. (1969). Sur un théorème de Lewis et la décomposition en facteurs premiers de la loi rectangulaire. *J. Appl. Probability* **6**, 177–185.

Wintner, A. (1934). On the addition of independent distributions. *Amer. J. Math.* **56**, 8–16.

Wintner, A. (1938). Asymptotic distributions and infinite convolutions (mimeographed). Edward Brothers, Ann Arbor, Michigan.

Wintner, A. (1947). The Fourier transforms of probability distributions (mimeographed). Baltimore. Maryland (published by the author).

Wolfe, S. J. (1971). On moments of infinitely divisible distribution functions. *Ann. Math. Statist.* **42**, 2036–2043.

Yosida, K. (1944). On the representation of functions by Fourier integrals. *Proc. Imp. Acad. (Tokyo)* **20**, 655–660.

Zygmund, A. (1947). A remark on characteristic functions. *Ann. Math. Statist.* **18**, 272–276.

Index

Probability and Mathematical Statistics

A Series of Monographs and Textbooks

---

1. Thomas Ferguson. Mathematical Statistics: A Decision Theoretic Approach. 1967
2. Howard Tucker. A Graduate Course in Probability. 1967
3. K. R. Parthasarathy. Probability Measures on Metric Spaces. 1967
4. P. Révész. The Laws of Large Numbers. 1968
5. H. P. McKean, Jr. Stochastic Integrals. 1969
6. B. V. Gnedenko, Yu. K. Belyayev, and A. D. Solovyev. Mathematical Methods of Reliability Theory. 1969
7. Demetrios A. Kappos. Probability Algebras and Stochastic Spaces. 1969
8. Ivan N. Pesin. Classical and Modern Integration Theories. 1970
9. S. Vajda. Probabilistic Programming. 1972
10. Sheldon M. Ross. Introduction to Probability Models. 1972
11. Robert B. Ash. Real Analysis and Probability. 1972
12. V. V. Fedorov. Theory of Optimal Experiments. 1972
13. K. V. Mardia. Statistics of Directional Data. 1972
14. H. Dym and H. P. McKean. Fourier Series and Integrals. 1972
15. Tatsuo Kawata. Fourier Analysis in Probability Theory. 1972
16. Fritz Oberhettinger. Fourier Transforms of Distributions and Their Inverses: A Collection of Tables. 1973
17. Paul Erdös and Joel Spencer. Probabilistic Methods in Combinatorics. 1973
18. K. Sarkadi and I. Vincze. Mathematical Methods of Statistical Quality Control. 1973
19. Michael R. Anderberg. Cluster Analysis for Applications. 1973
20. W. Hengartner and R. Theodorescu. Concentration Functions. 1973
21. Kai Lai Chung. A Course in Probability Theory, Second Edition. 1974
22. L. H. Koopmans. The Spectral Analysis of Time Series. 1974
23. L. E. Maistrov. Probability Theory: A Historical Sketch. 1974
24. William F. Stout. Almost Sure Convergence. 1974

25. E. J. McShane. Stochastic Calculus and Stochastic Models. 1974

26. Z. Govindarajulu. Sequential Statistical Procedures. 1975

27. Robert B. Ash and Melvin F. Gardner. Topics in Stochastic Processes. 1975

28. Avner Friedman. Stochastic Differential Equations and Applications, Volumes 1 and 2. 1975

29. Roger Cuppens. Decomposition of Multivariate Probabilities. 1975

In Preparation

Eugene Lukacs. Stochastic Convergence, Second Edition

Harry Dym and Henry P. McKean. Gaussian Processes: Complex Function Theory and the Inverse Spectral Method

A B C D E F G H I J